Making Everyday Electronics Work

A Do-It-Yourself Guide

Stan Gibilisco

New York   Chicago   San Francisco
Lisbon   London   Madrid   Mexico City
Milan   New Delhi   San Juan
Seoul   Singapore   Sydney   Toronto

Sponsoring Editor
Roger Stewart

Editorial Supervisor
Stephen M. Smith

Production Supervisor
Pamela A. Pelton

Acquisitions Coordinator
Amy Stonebraker

Project Manager
Nancy Dimitry, D&P Editorial Services

Copy Editors
Joe Cavanagh, Nancy Dimitry,
D&P Editorial Services

Proofreader
Don Dimitry, D&P Editorial Services

Art Director, Cover
Jeff Weeks

Composition
D&P Editorial Services

About the Author

Stan Gibilisco, an electronics engineer and mathematician, has authored multiple titles for the McGraw-Hill *Demystified* and *Know-It-All* series, along with numerous other technical books and dozens of magazine articles. His work has been published in several languages.

Contents

Introduction

Have you ever felt lost when choosing a home entertainment system, wondered how to test an electrical outlet to see if it's grounded, or puzzled over how to make the Wi-Fi in the front room reach the back? If so, read on!

Here's a question that people occasionally ask me, and that baffles me as much as it does them: "What is electricity, *really*?" I'll never forget the day my eighth-grade science teacher showed the class a celluloid "movie" of a lecture where a professor concluded by saying, "We learn about electricity not by knowing *what it is*, but by codifying *what it does*." I adopted that attitude as I set out to write this book as a how-to guide for nontechnical people who want to learn more about home and automotive electrical and electronic systems, and in particular, what "makes them tick" (or not).

You'll find out why things sometimes fail, and I'll offer you some solutions to common problems (and tips on how to avoid them). You'll find sidebars to clear up points of confusion and offer targeted bits of advice. In the last chapter, I'll suggest some projects for those of you who might like to try out some "off-the-wall" projects and experiments. In the back of the book, you'll find a glossary that defines common terms in easy-to-understand language.

I welcome your suggestions for future editions. Please visit me on the Web at **www.sciencewriter.net**.

Stan Gibilisco

CHAPTER 1

Let's Start with the Basics

Electronic devices and systems have changed people's lives more in the past century than all prior inventions and events did, going back to prehistoric times. If you don't believe me, wait until you have to live through a long power outage in the wake of a hurricane, earthquake, or wildfire! Sooner or later you'll start to wonder if the Stone Age might be about to come back for good. What's behind all these marvels that present such a tenuous barrier between comfort and chaos? Let's find out what makes them work.

Direct Current

All matter comprises countless tiny particles called *atoms*. Individual atoms are made up of smaller particles known as *protons*, *neutrons*, and *electrons*.

Protons and neutrons are smaller than any ordinary microscope can "see," and they have phenomenal density. A pebble-sized lump of compacted protons or neutrons would weigh so much that it would fall through the floors of your house and bore into the earth as if rock were butter.

In an atom, the protons and neutrons always exist in a "clump" called the *nucleus*. (Hydrogen in its most abundant form serves as the lone exception; its whole nucleus is only one proton, all alone.)

Electrons are much less dense than protons or neutrons, and they move a lot more. Electrons can "orbit" around a single nucleus, wander among many different nuclei, or hurtle freely through space.

Did You Know?

Protons and electrons carry equal and opposite *electric charge*. Scientists consider protons as *electrically positive*, and electrons as *electrically negative*. These *charge polarity* definitions came about as a coincidental result of observations made long ago in simple experiments.

An excess or deficiency of electrons on an object gives that object a *static electric charge*, also called an *electrostatic charge*. If an object contains more total electrons than total protons, then that object has a *net negative charge*. If an object contains fewer total electrons than total protons, then that object has a *net positive charge*.

When charged particles move, you observe an *electric current*. Usually the current-carrying particles, known as *charge carriers*, are electrons. However, any moving charged object, such as a proton, an atomic nucleus, or an electrified dust grain, can give rise to an electric current. In *direct current* (DC), the charge carriers always travel in the same general direction.

An *electrical conductor* is a substance in which the electrons can move easily, so you don't have any trouble producing an electric current. Silver is the best-known everyday electrical conductor. Copper and aluminum are also excellent electrical conductors. Iron, steel, and most other metals constitute fair to good conductors of electricity. An *electrical insulator* is a substance in which electrons won't pass from atom to atom under ordinary circumstances.

Did You Know?

Most gases, being poor conductors, make good insulators. Glass, dry wood, paper, and plastics are other examples of excellent electrical insulators. Pure water also makes a decent insulator, although it conducts some current when it contains dissolved minerals (such as you usually find in tap or well water).

Current can flow only if charge carriers are "pushed" or otherwise forced to move. The "push" can result from a buildup of electrostatic charge, or from a steady charge difference between two objects. When you have a positive charge pole (relatively fewer electrons) in one place and a negative charge pole (relatively more electrons) in another place not far away, an *electromotive force* (EMF) exists between the two charge poles. Electricians and engineers express this force, also known as *voltage* or *electrical potential*, in units called *volts* (symbolized V). You can say that a *potential difference* exists between the two charge poles. Once in a while, lay people refer to voltage as "electrical pressure."

Fact or Myth?

Has anyone ever told you that it's the current, not the voltage that makes an electrical system dangerous? This statement holds true in a literal sense, but it oversimplifies the real situation. In theory, high voltage all by itself can't harm anybody. However, deadly current can flow only when sufficient voltage exists to drive it. The current is directly responsible for electrocution, but a dangerous current can't flow without enough voltage to propel a lot of charge carriers through your body. If someone says that high voltage won't harm you but high current will, it's sort of like saying that if you jump off a cliff, the fall won't kill you but the impact will! It's just stupid semantics.

Warning! Even a moderate voltage can pose a deadly danger under some conditions. When you're working around anything that carries more than about 12 V (the voltage from a common automotive battery), you'd better give that thing the same respect as you would do if it were a campfire, a chain saw, a big dog, or anything else that would put a reasonable person on a state of alert.

When you work with DC, the current through an electrical component varies in direct proportion to the voltage across it, as long as the characteristics of the component don't change. If you double the voltage, the current doubles. If the voltage falls to 1/10 of its original value, so does the current. Figure 1-1 shows how the current varies as a function of the voltage through a component whose *electrical conductance* always stays the same. This simple *linear* (straight-line) relationship holds true only as long as the conductance remains constant. In some components, the conductance changes as the current varies. An electric light bulb is a good example. The conductance is lower when the filament carries a lot of current and glows white hot, as compared to when it carries only little current and hardly glows at all.

For Nerds Only

If you have my book *Electricity Experiments You Can Do at Home* (McGraw-Hill, 2010), you can find instructions for a test that demonstrates how the conductance of a lantern bulb varies depending on how much voltage you apply to it. Refer to the conclusion of Experiment DC-17 (pages 100 and 101).

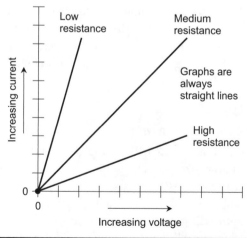

FIGURE **1-1**   When the voltage across a component increases but nothing else changes, the current increases in direct proportion to the voltage. If we graph the current versus the voltage, we get a straight line.

Warning! Before you work on any electrical appliance or system, unplug the appliance from the utility outlet and shut off the outlet's supply of electricity at the fuse or breaker box. These two precautions will minimize the danger to you in case something unexpected happens.

Nothing conducts electricity perfectly. Even the best conductors have a little bit of *resistance*, which you can define as an *impediment* to the flow of current (the opposite of conductance). Silver, copper, aluminum, and most other metals have excellent conductance, so they have low resistance. Some materials, such as carbon and silicon, have moderate resistance. Electrical insulators exhibit high resistance. Resistance and conductance vary in *inverse proportion* with respect to each other.

Engineers express and measure electrical resistance in units called *ohms*. In some texts, you'll find the word "ohm" or "ohms" symbolized by the uppercase Greek letter omega (Ω). As a component's resistance in ohms gets greater and greater, current has more and more trouble flowing through it, given a constant voltage across it. Conversely, as you reduce a component's resistance in ohms while applying a constant voltage across it, you get more current through it.

In an everyday electrical system with common appliances, such as lamps, television sets, computers, refrigerators, and the like, you'll always want to keep the resistance as low as possible. Resistance converts electrical energy into thermal energy (heat). This phenomenon is called *resistance loss* or *ohmic loss*. Except in devices such as space heaters or ovens whose primary purpose is to generate heat, ohmic loss represents energy going to waste.

Did You Know?

When current flows through a component that has a certain amount of resistance, the current gives rise to a voltage between its two end terminals. If the component's resistance remains constant, an increase in the current through it will produce a corresponding increase in the voltage across it. Remember, the voltage across a component varies in direct proportion to the current that flows through it, as long as the resistance doesn't change.

Quick Question, Quick Answer

- Imagine a "black box" (BB) connected to a battery. Suppose that the BB has a certain resistance. The voltage from the battery causes some current to flow through the BB. What will happen to the current if you double the battery voltage, and at the same time cut the resistance of the BB in half?
- The current will go up by a factor of 4. If the resistance of the BB didn't change, then doubling the voltage would double the current. If the voltage

across the BB didn't change, then cutting the resistance in half would double the current. But because both things happen at once (the voltage doubles and the resistance drops to half), the current gets multiplied by a factor of 2 twice over, which equals 4.

Alternating Current

The charge carriers (usually electrons) in *alternating current* (AC) don't keep moving in the same direction all the time. They reverse direction at regular intervals. In a household electric circuit, that time interval is 1/120 of a second (in the United States and some other countries) or 1/100 of a second (in most of the remaining countries). The current reverses twice, coming back to its original starting point, at intervals of 1/60 or 1/50 of a second, respectively. Therefore, you can say that the current goes through one complete *cycle* every 1/60 or 1/50 of a second.

You can portray an AC wave as a graph of voltage versus time, with voltage on the vertical axis and time on the horizontal axis. When you do that, you get a wave as shown in Fig. 1-2. You can portray a single AC cycle as any portion of the wave that lies between a fixed point and the corresponding point on the next repetition of the wave. Figure 1-2 shows two successive *wave crests* at which the wave reaches its maximum positive voltage. The time required for a single cycle to take place corresponds to the distance in the graph between any two adjacent crests. Figure 1-2

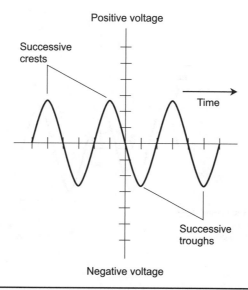

FIGURE 1-2 Successive crests or successive troughs can define a wave cycle.

also shows two successive *wave troughs* at which the wave attains its maximum negative voltage. As with crests, the time period for one cycle corresponds to the distance between any two successive troughs.

You can express one cycle of a wave by determining the time interval between any two adjacent points where the wave crosses the time axis (the horizontal axis) going up, or between any two adjacent points where the wave crosses the time axis going down. Because any point on the time axis indicates a voltage of zero at that instant, scientists call it a *zero point*. The zero points correspond to a momentary absence of voltage. Figure 1-3 shows two successive *positive-going zero points* and two successive *negative-going zero points*.

The *instantaneous voltage* of an AC wave is the voltage at some precise moment, or instant, in time. The instantaneous voltage of the AC from a utility outlet constantly varies, in contrast to the instantaneous voltage of the electricity from a battery, which remains constant as time passes (as long as the battery holds its charge).

When plotted as a graph with time on the horizontal axis and voltage on the vertical axis, a conventional AC utility wave always resembles the undulating curves that you see in Figs. 1-2 and 1-3. All of the electrical energy exists at a single, constant frequency. Any AC energy that exists entirely at a single frequency produces a characteristic graph called a *sine wave*. These days, electricians and engineers talk about units called *hertz* instead of cycles per second. They abbreviate hertz as Hz. The word "hertz" means the same thing as the expression "cycles per second."

Let's define the *positive peak voltage* of an AC wave as the maximum positive instantaneous voltage that the wave attains. Similarly, the *negative peak voltage*

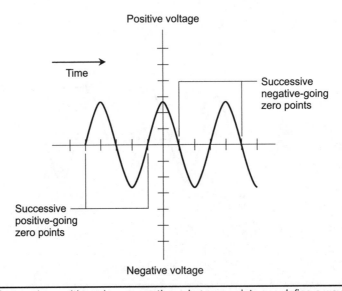

FIGURE 1-3 Successive positive-going or negative-going zero points can define a wave cycle.

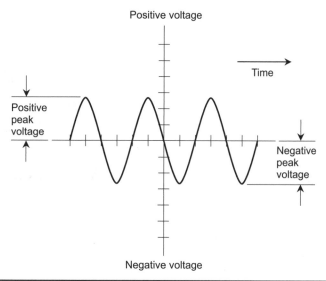

FIGURE 1-4   Positive peak and negative peak voltages for an AC wave.

equals the maximum negative instantaneous voltage. In household AC waves, the positive and negative peak voltages are equal and opposite as shown in Fig. 1-4.

Did You Know?

Neither the positive nor the negative peak voltage of an AC wave has anything to do with the frequency. You can change the frequency from a millionth of a hertz to millions of hertz, and keep the positive and negative peak voltages the same. Conversely, you can change the peak voltages from a millionth of a volt to a million volts and keep the frequency the same.

The *peak-to-peak voltage* of an AC wave equals the mathematical difference between the positive peak voltage and the negative peak voltage, taking polarity into account. If the positive and negative peak voltages have identical extent but opposite polarity (always the case in utility AC), then the peak-to-peak voltage equals twice the positive peak voltage, or −2 times the negative peak voltage. Figure 1-5 shows the peak-to-peak voltage in the graph of an AC sine wave. Peak-to-peak voltage doesn't need to have any polarity defined. In fact, polarity is irrelevant; the peak-to-peak voltage is simply a "plain old number."

In real-world AC systems, you won't have to worry about peak or peak-to-peak voltages very often. You'll be more concerned about the *effective voltage*. For an AC wave that is the voltage that a *pure DC source* (such as a battery) would have to generate to produce the same effect as the AC wave does in a component with a certain resistance. The most common way to express effective AC voltage is the *root-mean-square* (RMS) method. A computer could closely approximate the RMS

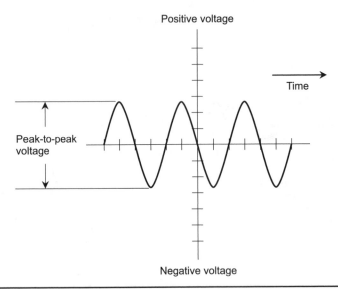

FIGURE 1-5 Peak-to-peak voltage for an AC wave.

value of a wave by squaring all of the instantaneous values at microsecond (or even nanosecond!) intervals, then averaging them, and finally taking the square root of the result. The calculation involves taking the square *root* of the *mean* of the *squares*; that's where the term "root-mean-square" comes from.

For Nerds Only

For a household AC wave or any other sine wave, the RMS voltage equals about 0.354 times the peak-to-peak voltage, and the peak-to-peak voltage equals about 2.83 times the RMS voltage.

Did You Know?

When an electrician says that a wall outlet supplies 117 V, she means to say that it puts out 117 root-mean-square volts (117 V RMS)—not 117 average volts, not 117 positive peak volts, not –117 negative peak volts, and not 117 peak-to-peak volts.

Quick Question, Quick Answer

- What's the peak-to-peak voltage of household AC that appears at your wall outlets at 117 V RMS?
- Multiply 117 by 2.83 and then round off the result to the nearest volt. You should get 331 V peak-to-peak.

Fact or Myth?

If anybody ever tells you that the RMS voltage of an AC wave is the same thing as its *average* voltage, don't believe it! The average voltage of a normal utility wave equals zero! That's because the wave goes positive and negative in exact "mirror images" from zero volts. You can see this fact by looking at any of the graphs in Figs. 1-2 through 1-5. Over time, the instantaneous voltages all average out to zero volts. But that's a whole lot different from having no electricity at all, as you know if you've ever gotten a shock from a household AC appliance carrying "zero average volts." The effective or RMS voltage is defined in an entirely different way than the average voltage, and the resulting quantities are almost always different—sometimes vastly different.

Electrochemical Cells and Batteries

The *electrochemical cells* sold in stores, and used in common devices, such as flashlights and portable headsets, produce about 1.5 V. You'll find them in sizes called AAA (very small), AA (small), C (medium), and D (large). When you combine two or more cells to increase the current or voltage (or both), you get a *battery*. Batteries produce voltages from about 1.5 V up to a few dozen volts.

Did You Know?

In the olden days when every type of electronic device, even those portable "walkie-talkies," had vacuum tubes, you could find some batteries that went over 100 V. One of the most common types was the same size as (but a lot heavier than) the cardboard tube that you get after you've used up a roll of paper towels. If you had one of those batteries in your hands right now, and if it was "fresh," you could get a deadly shock from it. I remember using one of those old things in the 1960s to light up a standard 15-watt utility bulb to almost full brilliance. Amazing!

In a *zinc-carbon cell*, the outer case is a thick foil made of zinc. It acts as the negative electrode. A carbon rod serves as the positive electrode. An internal chemical called the *electrolyte*, which gives the cell its energy, comprises a paste of manganese dioxide and carbon. Zinc-carbon cells are fairly cheap to produce, so they won't cost you much at the store. They work well at moderate temperatures, and in applications where the current drain ranges from moderate to high. They "lose their juice" if the temperature falls very far below the freezing point of water.

Alkaline cells contain granular zinc as the negative electrode, potassium hydroxide as the electrolyte, and a substance called a *polarizer* as the positive electrode. An alkaline cell can work at lower temperatures than a zinc-carbon cell can. You can expect an alkaline cell or battery to last a long time in a low-current electronic device, such as a portable calculator or electronic clock. Alkaline cells and batteries last longer in most situations than zinc-carbon cells or batteries do, but they cost more.

Silver-oxide cells are usually molded into button- or pill-like shapes. For this reason, they're often called *button cells* (although some other types of cells also have this shape, and are also called button cells). Silver-oxide cells can fit inside wristwatches and digital cameras. They come in various sizes and thicknesses, supply 1.5 V, and offer excellent energy storage capacity for their size and weight. Button cells can be stacked to make compact batteries that provide several volts as shown in Fig. 1-6.

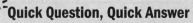

Quick Question, Quick Answer

- Suppose that you want to make a 9-V battery by stacking up multiple silver-oxide button cells. How many cells will you have to use? How can you make sure that the polarity is correct?
- A single silver-oxide cell produces 1.5 V, so you'll need six of them to make a 9-V battery. Connected in series (end to end, negative to positive, like the cars in a train or the links in a chain), the six cells produce 6×1.5 V, which equals 9 V. You must orient all the cells in the same direction. You should also pay attention to the polarity when using the battery. You'll see the cell faces labeled with either a plus sign (+) or a minus sign (−) to tell you the polarity

Mercury cells, also called *mercuric oxide cells*, have advantages similar to silver-oxide cells. They're manufactured in the same button-like shape. The main difference, often not of significance, is a somewhat lower voltage per cell: approximately 1.35 V instead of 1.5 V. In recent years, mercury cells and batteries have fallen from favor because mercury has been recognized as a toxic heavy metal that tends to accumulate in the environment (and in the vital organs of living things, like us). When you discard a mercury cell or battery, you must observe special precautions. In some locations, strict laws govern the disposal process.

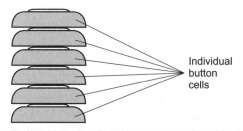

Individual button cells

Figure 1-6   You can stack up silver-oxide button cells to make a battery with higher voltage than the individual cells have.

Did You Know?

If you have cells or batteries that you think might contain mercury, call your local trash-removal department, and get instructions on how to dispose of the things. Don't simply throw them into a rubbish bin!

Lithium cells supply 1.5 V to 3.5 V, depending on the process used in their manufacture. These cells, like their silver-oxide cousins, can be stacked to make batteries. Lithium cells and batteries have superior shelf life, and they can last for years in very low-current applications. They have exceptional energy capacity per unit volume. Some engineers believe that lithium batteries will play a key role in the future of electric motorized personal transportation (electric bicycles, motorcycles, all-terrain vehicles, cars, boats, and snowmobiles) in the years and decades to come.

A *lead-acid cell* contains a liquid electrolyte of sulfuric acid, along with a lead negative electrode and a lead-dioxide positive electrode. You can connect several such cells in series (negative-to-positive at each connection) to get a battery that will provide you with useful power for several hours. Some lead-acid batteries contain an electrolyte thickened into a paste to reduce the danger of leakage. These components do a good job in consumer devices that require moderate current, such as notebook computers, tablet computers, electronic-book readers, and cell phones. They're also used in *uninterruptible power supplies* (UPSs) that can provide short-term emergency backup power for desktop computers.

Transistor batteries are miniature box-shaped things, roughly the size of a small pack of chewing gum, equipped with clamp-on terminals. They produce 9 V and contain six tiny zinc-carbon or alkaline cells in series. Transistor batteries are used in very-low-current electronic devices that operate on an intermittent (as opposed to continuous-duty) basis, such as wireless garage-door openers, appliance remotes, smoke detectors, carbon-monoxide detectors, and electronic calculators.

Lantern batteries are rather bulky and heavy, and they can deliver a fair amount of current. This type of battery usually contains multiple zinc-carbon or alkaline cells in a series-parallel combination, producing a net output of 6 V DC. One type of lantern battery has spring contacts on the top. The other type has thumbscrew terminals. A single lantern battery can keep a small, low-voltage DC incandescent bulb lit for quite a while, and a set of *light-emitting-diode* (LED) lamps aglow for days. A pair of them in series can deliver enough power to operate a citizens band (CB) radio set or a low-power ham radio set for several hours.

Nickel-cadmium (NICAD) batteries are sometimes found in older portable electronic devices. They come in box-shaped packages that insert directly into the equipment to form part of the case. An example is the battery pack for a handheld communications transceiver for amateur, CB, police, or military use. In recent years, *nickel-metal-hydride (NiMH) batteries* have supplanted NICAD batteries. The NiMH chemistry doesn't contain cadmium, which, like mercury, acts as a toxin when it gets into the environment.

Fact or Myth?

Have you heard that nickel-based batteries exhibit a bothersome characteristic called *memory* or *memory drain*? According to popular wisdom, if you use such a device repeatedly, and if you discharge it to approximately the same extent with every cycle, it will lose its ability to hold a charge for as long as it should. Some engineers say that this phenomenon hardly ever occurs, but I've seen it happen with NICADs. You can sometimes "cure" a nickel-based cell or battery of this problem by discharging it almost completely, recharging it, discharging it almost completely again, and repeating the cycle numerous times. But hopefully you'll never have to bother with that process. These days, most devices use lithium batteries instead of nickel-based ones, and lithium batteries don't get the "memory drain disease."

Did You Know?

When a NICAD or NiMH cell or battery has discharged almost all the way, you should stop using it and fully recharge it as soon as possible. Otherwise, it might permanently lose its energy-storage capacity as a result of cell-polarity reversal.

All rechargeable cells and batteries work best if you charge them with *trickle-charge* or *slow-charge* devices. Some of these devices plug into the *Universal Serial Bus* (USB) ports of computers, but smaller computers such as notebooks might not provide enough power to fully charge the battery. You'll always get the best results if you use a charger that plugs into an AC wall outlet. Always use a charger designed specifically for the type of cell or battery that you want to charge. So-called *quick chargers* are available, but some of them force excessive current through a cell or battery, causing permanent damage. You should allow several hours for the battery recharging process. My tablet computers take four to six hours to go from 10 percent to 95 percent of full charge.

For Nerds Only

If you find a cell or battery that doesn't have any labeling to tell you which end is positive and/or which end is negative, you can (and should) test it for polarity with a *volt-ohm-milliammeter* (VOM), also called a *multimeter*. You can buy a VOM for a few dollars at hardware stores or hobby electronics stores such as Radio Shack. You'll learn more about this neat little tool in the next chapter.

Power Supplies

An *electrical power supply* changes utility AC to pure DC, serving as an alternative to batteries for electronic devices. Figure 1-7 illustrates the major components of a power supply that converts 117 V RMS AC to constant-voltage DC.

The *fuse* or *breaker* protects the power supply, and the equipment connected to it, from damage in case of a malfunction such as a short circuit. Fuses are available in two types: *quick-break* and *slow-blow*. A quick-break fuse contains a straight length of wire or a metal strip. A slow-blow fuse contains a spring along with the wire or strip. You should always replace blown-out fuses with new ones of the same type. Quick-break fuses in slow-blow situations might burn out needlessly. Slow-blow fuses in quick-break situations might not adequately protect the equipment.

A breaker performs the same function as a fuse, but it's easier to reset. Instead of physically removing the fuse, finding a new one, and making sure that it's the right type and then installing it, you have only to switch off the power supply, wait a moment, and then press a button or flip a switch. Some breakers reset automatically when the equipment has remained powered down (switched off) for a few minutes.

Warning! Never replace a fuse with a larger-capacity unit to overcome the inconvenience of repeated blowing-out. Find the cause of the trouble, and repair the equipment as needed. The "penny in the fuse box" scheme can endanger you and your equipment, and it increases the risk of fire in the event of a short circuit.

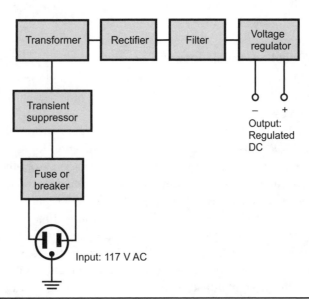

Figure 1-7 Block diagram showing the major components of a well-engineered power supply for converting utility AC to DC that an electronic device can use.

The AC on the utility line is supposed to be a sine wave at 60 Hz, without any flaws or distortions whatsoever. But in fact, it's far from "pure." If you look at the AC waveform on a high-quality laboratory oscilloscope, you'll occasionally see *voltage spikes*, known as *transients* that greatly exceed the positive or negative peak waveform voltage. Transients can result from sudden changes in the *load* (the amount of power demanded by all the appliances combined) in a utility circuit. Lightning is notorious for causing destructive transients. A single thundershower can produce transients throughout a neighborhood or small town. Unless you take measures to suppress them, transients can destroy some components in a power supply. Transients can also interfere with the operation of sensitive electronic equipment, such as computers or microcomputer-controlled appliances.

You can find commercially made *transient suppressors* in most hardware stores and large department stores. These devices, often mistakenly called "surge protectors," use specialized semiconductor-based components to prevent sudden voltage spikes from reaching levels where they can cause problems. These devices are sometimes rated in *joules*, indicating the severity of the transients they can protect against. The higher the "joule rating," the better.

Did You Know?

You should use transient suppressors with all sensitive electronic equipment including computers, hi-fi stereo systems, and television sets (especially those expensive big-flat-screen ones). In the event of a thunderstorm, the best way to protect such equipment is to physically unplug every single appliance from its wall outlet until the storm has passed.

Quick Question, Quick Answer

- Will a transient suppressor work if it's designed for a three-wire electrical system but the "third prong" of the plug that goes into the wall outlet has been defeated or cut off?
- No. In order to properly function, a transient suppressor requires a good *electrical ground* connection. Such a connection must ultimately lead to a ground rod driven into the earth. Every residence or other building should have such a *ground rod*, preferably at the point where the utility line enters the structure. Transients can get directed away from sensitive equipment only when a current path exists for discharge to a good electrical ground. In order to guarantee such a connection, the building's wiring must have three-wire outlets, and the "third holes" in those outlets must lead to a well-designed electrical ground. If you have any doubts about the ground system in your house or business, have it checked out by a competent electrician.

A *transformer* converts, or *transforms*, an AC sine wave with a given voltage to another AC wave with the same frequency but a different voltage. A typical transformer contains wires wound on a special form called a *core* made of *laminated iron* (thin slabs of iron glued together). The wires wrap around the core to make windings called the *primary* and the *secondary*. The primary is the winding to which you apply the electricity whose voltage you want to change. The secondary is the winding from which you take the electricity after its voltage has changed.

- In a *step-down transformer*, the secondary has fewer turns than the primary has. The voltage across the secondary (the output voltage) is, therefore, lower than the voltage across the primary (the input voltage).
- In a *step-up transformer*, the secondary has more turns than the primary has. The output voltage is, therefore, higher than the input voltage.

Small step-down transformers are used in simple power supplies and battery chargers for things like computers and radios. Medium-sized transformers find application in high-current or high-voltage power supplies for things like amateur ("ham") radio amplifiers and those big, old-fashioned, picture-tube-type television sets. Large step-down transformers provide the utility power that people consume in homes and businesses. The most massive transformers can get as big as a house; they serve power-generating plants and transmission stations. Some of these electrical behemoths step the voltage down; others step it up.

For Nerds Only

An AC wave can pass through a transformer, but DC can't. If the input wave to a transformer contains some DC "on top of" the AC, then the *DC component* will not appear in the output. Conversely, if you force DC through the secondary of a transformer, that component won't get "fed back" to the primary circuit. Engineers sometimes take advantage of this DC-blocking property when they use transformers in specialized electronic devices.

A *rectifier* converts AC to *pulsating DC*, usually by means of one or more heavy-duty *semiconductor diodes* following a power transformer. The simplest type, called a *half-wave rectifier*, uses one diode to cut off half of the AC cycle (either the positive half or the negative half). Half-wave rectification works okay in a power supply that never has to deliver much current, or when the voltage can vary without affecting the behavior of the equipment connected to it.

A more sophisticated rectifier circuit takes advantage of both halves of the AC cycle, rather than only the positive half or the negative half, to obtain pulsating DC. In most applications, the *full-wave rectifier* offers the best method for converting AC to DC. This type of rectifier uses four diodes. The increased circuit complexity rarely amounts to much in terms of cost (most rectifier diodes are cheap), but it makes a big difference when a power supply must deliver a lot of current.

Most electronic equipment requires something better than the pulsating DC that comes straight out of a rectifier circuit. A *filter* can minimize the roughness called *ripple* that always appears in the DC from a rectifier. The simplest power-supply filter comprises a large-value capacitor, connected in parallel with (that is, across) the rectifier output, between the positive and negative terminals. An *electrolytic capacitor* works well in this role. It's a *polarized* component, meaning that you must connect it in a certain direction, just as you would do with a battery.

Did You Know?

Filter capacitors work by "trying" to maintain the DC voltage at its peak level as the output of the rectifier goes through its pulsations. This task is easier to do in the output of a full-wave rectifier than in the output of a half-wave rectifier. With a full-wave rectifier receiving a 60-Hz AC electrical input, the ripple frequency equals 120 Hz. With a half-wave rectifier, the ripple frequency equals 60 Hz.

Fact or Myth?

Has anyone ever told you that an electrolytic capacitor can "blow up" if you put too much voltage across it, or if you connect it with the polarity reversed? Did you think they were joking? Well, they weren't! If you mistreat an electrolytic capacitor, it can explode like a firecracker. I've seen it happen, and it's dangerous. Always double-check the polarities and voltage ratings of all your filter capacitors as you build a power supply (if you decide to do that), and check everything again before you apply power to the system for the first time. The capacitors should be rated at several times the actual power supply output voltage.

For Nerds Only

When a power supply operates without any load (in other words, with no appliances connected to it whatsoever), the voltage across the filter capacitors holds steady near the peak rectifier output voltage, not the RMS output voltage. Therefore, the DC voltage across the capacitors exceeds—sometimes greatly—the RMS pulsating voltage that charges them. This dramatic voltage increase explains why, whenever you build a DC power-supply filter, you should use capacitors rated to handle several times the rectifier's effective DC output voltage. It's better to err on the side of "overengineering" than to have a filter capacitor rupture or explode.

Whatever else a power supply does, it won't serve your needs if its voltage varies significantly when the load changes. You'll want your electronic devices to "see" the power supply exactly as they'd "see" a heavy-duty battery that provides pure DC at a constant voltage. That's why any good power supply must include a *voltage regulator*.

If you connect a specialized device called a *Zener diode* along with another component called a *resistor* in the output of a power supply, the combination will limit the output voltage to whatever value you choose. The Zener diode must have an adequate power rating to prevent it from burning out. The limiting voltage depends on the particular Zener diode that you use. You can find Zener diodes to fit any reasonable power-supply voltage. When you need a power supply to deliver high current, you can use a *power transistor* along with a Zener diode to regulate the voltage.

You can find prepackaged voltage regulators in *integrated-circuit* (IC) form. Such an IC, sometimes along with some external components, should be installed in the power-supply circuit at the output of the filter.

Grounding and Glitches

The best electrical ground for a power supply is the "third wire" ground provided in up-to-date AC utility circuits. The "third hole" (the bottom hole in an AC outlet, shaped like an uppercase English letter D turned on its side) should connect directly to a wire that runs to a *ground rod* driven into the earth at the point where the electrical wiring enters the building.

In old buildings, *two-wire AC systems* are common. They have only two slots in the utility outlets. Some of these systems employ reasonable grounding by means of *polarization*, where one slot is longer than the other, and the longer slot goes to electrical ground. But that method never works as well as a true *three-wire AC system*, in which the ground connection remains independent of both outlet slots.

Did You Know?
Unfortunately, the presence of a three-wire or polarized outlet system doesn't guarantee that an appliance connected to an outlet will be well-grounded. If the appliance design is faulty, or if the people who installed the electrical system didn't ground the "third slot," a power supply can deliver a dangerous voltage to external metal surfaces of appliances and electronic devices. This situation can pose an electrocution hazard, and can also hinder the performance of electronic equipment.

Warning!   All metal chassis and exposed metal surfaces of AC power supplies should be connected to the grounded wire of a three-wire electrical cord. *Never* defeat or cut off the "third prong" of the plug. As mentioned before (but it bears repeating), you should find out whether or not the electrical system in the building was properly installed so that you don't labor under the illusion that your system has a good ground when it really does not. If you have any doubts about this issue, hire a professional electrician to perform a complete inspection of the system. Then, if the system fails to "meet code," get the work done as necessary to make it good. Don't wait for disaster to strike!

To prevent power "glitches" from causing trouble such as computer data loss, you can use an *uninterruptible power supply* (UPS) such as the one diagrammed in Fig. 1-8. Under normal conditions, the equipment gets its power through the transient suppressor, the transformer, and the *AC regulator*. The transient suppressor gets rid of potentially destructive voltage "spikes." The AC regulator eliminates *surges* and *dips* in the utility power. A small current through the rectifier and filter maintains the battery (usually a lead-acid type) in a fully charged state.

For Nerds Only

The term *surge* refers to the initial high current drawn by a cheap or poorly designed power supply when it's first switched on with a load connected, or to a momentary increase in power-line voltage that lasts longer than a transient but is less intense. The term *dip* refers to a momentary decrease in the power-line voltage, something like an "upside-down surge," that can occur when a large appliance first comes on. You've probably noticed these fluctuations as momentary increases or decreases in the brilliance of an old-fashioned incandescent bulb when some heavy appliance such as a washing machine or refrigerator starts up.

If a particularly drawn-out dip or power failure occurs, the UPS takes care of it for a few minutes. An *interrupt signal* causes the switch to disconnect the equipment from the regulator and connect it to the *power inverter*, which converts the battery DC output to AC at the normal utility voltage. Then the battery starts to discharge. Its capacity should be sufficient to last long enough to allow for proper system shutdown. Once you have shut down all of the equipment connected to the UPS, you can switch off the UPS. Hopefully, if the power outage lasts for a long time, you'll have some sort of emergency system in place, such as a gasoline-powered or propane-powered generator (or, if you're into high-tech stuff, a solar-powered backup). When utility power returns to normal, the battery starts to charge up again, whether the UPS is switched on or not.

Warning!   All power supplies (even the smallest ones, such as the little things that charge your tablet computer's battery) can pose a deadly danger whenever they're plugged into a wall outlet. Usually you're safe as soon as you unplug the power supply. But don't bet your life on that assumption! Some power supplies, and the circuits connected to them, can retain lethal voltages at exposed terminal points as a result of filter capacitors holding their charge, even after the entire system has been switched off, unplugged, and left unattended for quite a while. If you have any doubt about your ability to safely build, modify, repair, or otherwise work with the internal circuits of a power supply, leave the task to a professional technician.

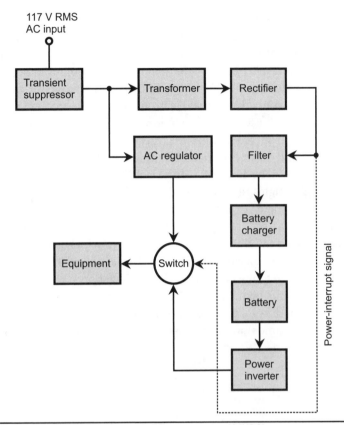

FIGURE 1-8   Block diagram of an uninterruptible power supply (UPS).

Magnetic Force

As children, we discovered that magnets "stick" to certain metals. Iron, nickel, a few other elements, and alloys or solid mixtures containing any of them constitute *ferromagnetic materials*. Magnets exert force on these metals. Magnets do not exert force on other metals unless those metals carry electric currents. Electrically insulating substances never "attract magnets" under normal conditions.

When you bring a magnet near a piece of ferromagnetic material, the atoms in the material line up to some extent, temporarily magnetizing the sample. This alignment produces a *magnetic force* between the atoms of the sample and the atoms in the magnet. Every atom acts as a tiny magnet; when they act in concert with one another, the whole sample behaves as a single large magnet. Magnets always "stick" to samples of ferromagnetic material.

If you place two magnets near each other, you observe a stronger magnetic force than you see when you bring either magnet near a sample of unmagnetized ferromagnetic material (an iron nail, say). The mutual force between two rod-shaped or bar-shaped magnets manifests as attraction if you bring two opposite poles close together (north-near-south or south-near-north) and repulsion if you bring two like poles into proximity (north-near-north or south-near-south). Either way, the force increases as the distance between the ends of the magnets decreases.

Whenever the atoms in a sample of ferromagnetic material align to any extent rather than existing in a random orientation, a "region of influence" called a *magnetic field* surrounds the sample. A magnetic field can also result from the motion of electric charge carriers. In a wire, electrons move in incremental "hops" along the conductor from atom to atom. In a permanent magnet, the movement of orbiting electrons occurs in such a manner that an *effective current* arises.

Physicists and engineers describe magnetic fields in terms of *flux lines*, also called *lines of flux*. The intensity of the field depends on the number of flux lines passing at right angles through a region having a certain cross-sectional area, such as a square centimeter or a square meter. The flux lines aren't material things, of course, but you can see their effects by doing a simple experiment.

Have you seen the classical demonstration in which iron filings lie on a sheet of paper, and then the experimenter holds a permanent magnet underneath the sheet? The filings arrange themselves in a pattern that shows, roughly, the "shape" of the magnetic field in the vicinity of the magnet. A bar magnet has a field whose lines of flux exhibit a characteristic pattern, as shown in Fig. 1-9.

Another experiment involves passing a current-carrying wire vertically through a sheet of paper oriented horizontally. The iron filings bunch up in circles centered

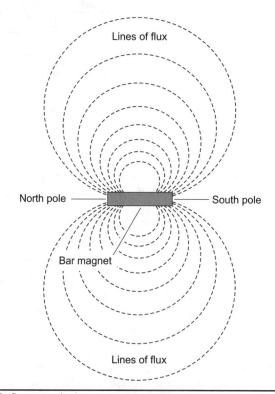

Figure 1-9   Magnetic flux around a bar magnet.

at the point where the wire passes through the paper. This experiment shows that the lines of flux around a straight, current-carrying wire form concentric circles in any plane passing through the wire at a right angle. The center of every "flux circle" lies on the wire, which serves as the path along which the charge carriers move (Fig. 1-10). A magnetic field has a specific orientation at any point near a current-carrying wire or a permanent magnet. At any point, the magnetic flux lines always run parallel with the direction of the magnetic field's "flow."

Did You Know?

Scientists consider any magnetic field to begin, or originate, at a *north pole*, and to end, or terminate, at a *south pole*. These poles don't correspond to the earth's magnetic poles, however. They're the opposite! The earth's north magnetic pole is actually a magnetic south pole because it attracts the north poles of compasses such as the ones hikers and hunters use. Similarly, the earth's south magnetic pole is really a magnetic north pole because it attracts the south poles of compasses.

Quick Question, Quick Answer

- How do the flux lines in a magnetic field differ from the flux lines in an electric field, such as the sort that surrounds an electrically charged particle?
- A charged particle hovering all by itself in space produces *electric flux lines* that aren't closed; they radiate away from the particle in all directions, like infinitely long spikes. But magnetic flux lines always form closed loops. In the vicinity of a magnet, you can always find a starting point (the north pole) and an ending point (the south pole). You can't have a magnetic north or south pole anywhere all by itself; it must have a "mate" of the opposite polarity nearby.

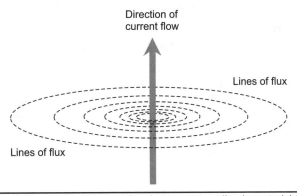

FIGURE 1-10   Magnetic flux produced by an electric current traveling in a straight line.

You might suppose that the magnetic field around a current-carrying wire, such as the one shown in Fig. 1-10, arises from a single, isolated magnetic pole. Or, you might imagine that no magnetic poles exist at all! The concentric flux circles don't seem to originate or terminate anywhere. You can get around this problem by means of a mind game. You can "invent" an originating point and a terminating point anywhere you want on one of the flux circles, thereby defining a pair of opposite magnetic poles in close proximity.

In a magnetic field, the lines of flux always connect the two magnetic poles. Some flux lines appear straight in a local sense, but in the larger sense, they always form curves. The greatest magnetic field strength around a bar magnet occurs near the poles, where the flux lines converge or diverge. As you move away from the poles, the magnetic field grows less intense. Around a current-carrying wire, the greatest field strength exists near the wire, and the intensity diminishes as you move away from the wire.

In theory, the flux field around any magnet, or around any current-carrying wire, extends into space indefinitely. In practice, the effects "wear off" at a certain distance from any magnet or wire because the field simply gets too weak to influence anything in the real world.

Did You Know?

Magnetic fields can arise from the motion of electrically charged subatomic particles through space, as well as from the motion of charge carriers through a conductor. The sun constantly ejects protons (which carry a positive charge) and electrons (which carry a negative charge). These particles produce electrical currents as they travel through space. The currents produce vast, fluctuating magnetic fields, just as a steady current in a wire generates a small-scale, stable magnetic field in a situation of the sort shown in Fig. 1-10. Conversely, fluctuating or unstable magnetic fields induce electric currents in electrical conductors, such as utility wires and radio antennas.

Fact or Myth?

We've all heard news reports from time to time, warning us that an eruption has taken place on the sun, and that we should prepare for possible disruptions to our communications or utility infrastructures. Are these warnings exaggerated? To some extent, maybe so; but when a *solar flare* occurs, the sun ejects far more charged particles than usual. As these particles approach the earth, their magnetic fields, working together, disrupt our planet's magnetic field, spawning a *geomagnetic storm*. Such an event can temporarily wipe out "shortwave radio" communications. In addition, people who live at high latitudes witness *aurora borealis* ("northern lights") and *aurora australis* ("southern lights") at night. If a big enough geomagnetic storm occurs, it can interfere with wire communications and electric power transmission at the surface. No one really knows (as of this writing, anyway) whether or not a massive solar flare will ever cause a worldwide power blackout lasting for years. But people have already witnessed dramatic effects. All the way back in the year 1859, a geomagnetic storm produced a so-called *electromagnetic pulse* (EMP) strong enough to generate currents in telegraph wires that set some stations on fire.

Electromagnets

The motion of electrical charge carriers always produces a magnetic field. This field can reach considerable intensity in a tightly coiled wire having a lot of turns and carrying a lot of current. When you place a ferromagnetic rod called a *core* inside a wire coil, as shown in Fig. 1-11, the magnetic lines of flux concentrate in the core, making the core sample into a powerful temporary magnet: an *electromagnet*.

Most electromagnets have rod-shaped cores. When you wind a wire into a coil around a rod-shaped object, you get a *solenoid*. A solenoid's length-to-diameter ratio can vary from extremely low (like a fat pellet) to extremely high (like a thin stick). Regardless of the length-to-diameter ratio, however, the flux produced by current in the solenoid's coil temporarily magnetizes the core that runs through it.

You can build a DC electromagnet by wrapping insulated wire around a large iron bolt. You can find these items in any good hardware store. You should test the

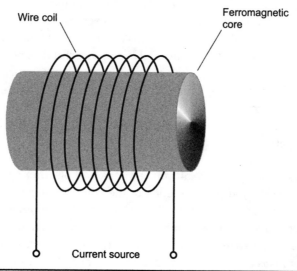

Wire coil

Ferromagnetic core

Current source

Figure 1-11 A simple electromagnet.

bolt for ferromagnetic properties while you're still in the store, if possible. (If a permanent magnet "sticks" to the bolt, then the bolt is ferromagnetic.) Ideally, the bolt should measure at least 3/8 inch (approximately 1 centimeter) in diameter and at least 6 inches (roughly 15 centimeters) in length. You must use insulated wire, preferably made of solid, soft copper. Don't use bare wire!

Wind the wire at least 100 times around the bolt. You can layer two or more windings if you like, as long as the wire always keeps going around in the same direction. Secure the wire in place with electrical or duct tape. A large "lantern battery" can provide plenty of DC to operate the electromagnet. You can connect two or more such batteries in parallel to increase the current delivery. Never leave the coil connected to the battery for more than a few seconds at a time.

Warning! Don't even think about using an automotive battery for the above-described experiment! The near short-circuit produced by an electromagnet can cause the acid from this type of battery to boil out, resulting in serious injury to you, not to mention possible damage to objects in the vicinity.

All DC electromagnets have well-defined north and south poles, exactly as permanent magnets have. However, an electromagnet can, at least in theory, get much stronger than any permanent magnet. The magnetic field exists only as long as the coil carries current. When you remove the power source, the magnetic field nearly vanishes. A small amount of *residual magnetism* remains in the core after current stops flowing in the coil, but this field is usually weak.

Some commercially manufactured electromagnets operate from 60-Hz utility AC. These magnets "stick" to ferromagnetic objects. The polarity of the field reverses every time the current reverses, producing 120 magnetic-field "pulses" every second, assuming a 60-Hz AC line frequency. The instantaneous intensity of the magnetic field varies along with the AC cycle, reaching alternating-polarity peaks at 1/120-second intervals and nulls of zero intensity at 1/120-second intervals.

Some electromagnets produce fields so powerful that no human can pull them apart if they get "stuck" together, and no human can push them all the way together against their mutual repulsive force. Industrial workers sometimes use huge electromagnets to carry heavy pieces of scrap iron or steel from place to place. Other electromagnets can provide sufficient repulsion to suspend one object above another, an effect known as *magnetic levitation*.

Warning! Do you think you can make an electromagnet "super powerful" if you plug the ends of the coil directly into an AC utility outlet? In theory, you can, but don't try it! You'll short out your house wiring, expose yourself to the risk of electrocution, expose your house to the risk of fire, and probably cause a fuse to blow or a circuit breaker to open, cutting power to the device anyway. Some buildings lack proper fuses or breakers, and shorting out one of those systems can lead quickly to disaster. If you want to build a safe AC electromagnet, my book *Electricity Experiments You Can Do at Home* (McGraw-Hill, 2010) offers instructions for doing it.

Magnetic Devices

Electrical relays, bell ringers, electric hammers, and other mechanical devices make use of solenoids. Sophisticated electromagnets, sometimes in conjunction with permanent magnets, allow engineers to construct motors, meters, generators, and other electromechanical devices. Let's look at a few examples.

Figure 1-12 illustrates a *bell ringer*, also called a *chime*. The ferromagnetic core has a hollow region in the center along its axis, through which a steel rod, called the *hammer*, passes. The coil has many turns of wire, so the electromagnet produces a strong field if high current passes through the coil. When no current flows in the coil, gravity holds the rod down so that it rests on the base plate. When a pulse of current passes through the coil, the rod jumps up and hits the ringer plate.

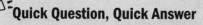

Quick Question, Quick Answer

- In a chime, such as the one shown in Fig. 1-12, the magnetic field "wants" the ends of the rod, which has the same length as the core, to align with the ends of the core. Why doesn't the rod stop at that point? Why does it continue on up to hit the ringer?
- The rod's upward momentum makes it fly through the core and keep going for a while, even as the magnetic field from the solenoid "tries" to pull it back. The rod travels far enough to hit the ringer and then falls back, allowing the ringer to reverberate.

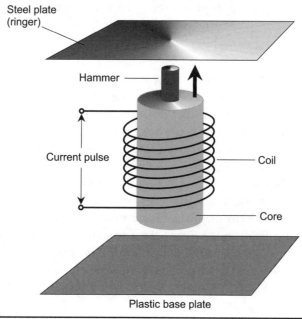

FIGURE 1-12   A bell ringer, also known as a chime.

You can't always locate switches near the devices they control. For example, imagine that you want to switch a communications system between two different antennas from a control point a few hundred meters away. Wireless antenna systems carry high-frequency AC (the radio signals) that must remain within certain parts of the circuit. You can't let those signals follow control wires that go to a simple switch; doing that would interfere with the workings of the antenna system. A *relay* makes use of a solenoid to allow remote-control switching in a situation of that sort.

Figure 1-13 illustrates a simple relay. A movable lever, called the *armature*, is held to one side (upward in this diagram) by a flexible, "springy strip" of metal or plastic when no current flows through the coil. Under these conditions, terminal X connects to terminal Y, but X does not contact terminal Z. When a sufficient current flows in the coil, the armature moves to the other side (downward in this illustration), disconnecting X from Y, and connecting X to Z.

A *normally closed relay* completes the circuit when no current flows in the coil, and breaks the circuit when coil current flows. ("Normal" in this sense means the absence of coil current.) A *normally open relay* does the opposite, completing the circuit when coil current flows, and breaking the circuit when coil current does not flow. The relay shown in Fig. 1-13 can function as a normally open or normally closed switch, depending on which contacts you select. It can also switch a single line between two different circuits.

Did You Know?

These days, engineers install relays in circuits and systems that must handle massive current or high voltage (or both). In applications where the current and voltage remain low to moderate, electronic semiconductor switches, which have no moving parts, offer better performance and reliability than relays.

In a *DC motor*, you connect a source of electricity to a set of coils that produce small-scale, but nevertheless powerful, magnetic fields. The attraction of opposite poles, and the repulsion of like poles, is manipulated so that a constant *torque* (rotational force) results inside the device. As the coil current increases, so does the torque that the motor can produce, and so does the energy it takes to operate the motor at a constant speed.

Figure 1-14 illustrates the functional aspects of a DC motor. The *armature coil* rotates along with the motor shaft. A pair of *field coils* remains stationary. The field coils function as electromagnets. (Some motors use a pair of permanent magnets instead of the field coils.) Every time the shaft completes half a rotation, the *commutator* reverses the current direction in the armature coil so that the shaft's torque keeps going in the same direction. The shaft's *angular* (rotational) *momentum* carries it around so that it doesn't stop at those instants in time when the current reverses.

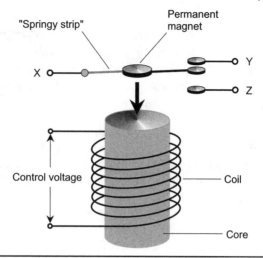

FIGURE 1-13 Simplified drawing of a relay.

The construction of an *electric generator* resembles the construction of an electric motor, although the two devices function in the opposite sense. A motor constitutes an *electromechanical transducer* because it converts electrical energy to mechanical motion. You might call a generator a specialized *mechanoelectrical transducer* (although I've never heard anybody use that term).

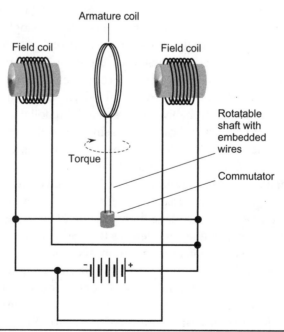

FIGURE 1-14 Simplified drawing of a DC motor.

A basic electric generator produces AC when a coil rotates in a strong magnetic field. You can drive the shaft with a gasoline-fueled engine, a turbine, or some other source of mechanical energy. Some generators employ commutators to produce pulsating DC output, which you can filter to obtain pure DC for use with precision equipment, just as you would do with the pulsating DC from an AC power supply.

Did You Know?

Some generators can operate as motors, and some motors can operate as generators. The experts call such devices *motor-generators*.

Semiconductors

During the 1960s, *semiconductor materials* acquired a dominating role in consumer electronic devices of all kinds. The term "semiconductor" arises from the fact that the substance's conductivity can be controlled to generate, amplify, modify, mix, rectify, and switch electrical currents or electronic signals. Various mixtures of elements and compounds can function as semiconductors. The two most common semiconductor media are based on the element *silicon* (Si) or a compound of *gallium* and *arsenic* known as *gallium arsenide* (GaAs).

In some semiconductor-based devices (also sometimes called *solid-state devices*), the power supply can be modest indeed, comprising a couple of 1.5-volt size AA or AAA "flashlight" cells or a 9-volt "transistor battery." A single *integrated circuit* (IC or *chip*), smaller than your thumbnail, can do the work of thousands of discrete electronic components, such as diodes, transistors, capacitors, and resistors. You'll find an excellent example of IC technology in any computer. In 1950, a personal computer (if such a thing had existed) would have occupied a large building, required thousands of watts to operate, and probably cost over a million dollars. Today you can buy one for a few hundred dollars and carry it in a small portfolio jacket.

Silicon is widely used in diodes, transistors, and ICs. Other substances, called *impurities* or *dopants*, are added to the silicon to give it the desired properties. If, on the other hand, you hear about "gasfets" and "gas ICs," you're hearing about GaAs technology. Gallium arsenide works better than silicon in several ways. A GaAs device needs less voltage than an equivalent Si device does. In addition, GaAs will function at higher frequencies than Si will. GaAs devices are relatively immune to the effects of ionizing radiation, such as x rays and gamma rays. GaAs devices are used in light-emitting diodes (LEDs), infrared-emitting diodes (IREDs), laser diodes, visible-light and infrared (IR) detectors, ultra-high-frequency (UHF) amplifying devices, and a variety of computer chips. The primary disadvantage of GaAs is the fact that it costs more than silicon to fabricate into semiconductor components.

Elemental *selenium* (Se) exhibits resistance that varies depending on the intensity of visible-light, infrared (IR) radiation, or ultraviolet (UV) radiation that falls on it. All semiconductor materials exhibit this property to some degree, but Se is especially affected. For this reason, Se makes excellent solar photocells and solar cells. This material is also used in certain types of rectifiers. Perhaps the main advantage of Se over Si is the fact that Se-based devices withstand power-line transients better than Si-based devices do.

Pure *germanium* (Ge) constitutes a rather poor electrical conductor, but it becomes a semiconductor when impurities are added. This substance was used extensively in the early years of semiconductor technology. Some diodes and transistors still use Ge, but it's pretty much been replaced by Si. One big problem with Ge-based technologies is their sensitivity to heat. Technicians must take extreme care when soldering the leads of a Ge component, so that the heat from the soldering instrument doesn't conduct through the wire leads and destroy the semiconducting properties of the Ge inside.

Some oxides of metals have properties that make them useful for semiconductor devices. When someone tells you about *MOS* (pronounced "moss") or *CMOS* (pronounced "sea-moss") technology, you're hearing about *metal-oxide semiconductor* and *complementary metal-oxide semiconductor* devices, respectively. Chips made from these materials demand so little power that the battery in a MOS-based or CMOS-based portable electronic device lasts almost as long as it would just sitting on the shelf without being put to any use at all. Devices with MOS and CMOS chips work fast, a property that makes them useful at high frequencies, allowing computers to perform many millions of calculations per second. In ICs, MOS and CMOS technology also allows for high *component density*: a large number (sometimes millions) of discrete diodes, transistors, capacitors, and resistors on a single chip.

Fact or Myth?

Some people might tell you that an MOS or CMOS device can be permanently ruined by an action as simple as picking it up and looking at it. Are they telling the truth? How could such a thing happen? Well, those people aren't lying or exaggerating. The main problem with MOS and CMOS technology arises from the fact that such devices are easily damaged by electrical discharges that can occur as a result of the accumulation of charge carriers somewhere. Even the slightest electrostatic buildup ("static electricity") on your fingers can abruptly "zero itself out" through the internal circuits of a MOS or CMOS device and destroy some of the microscopic components. You must always use care when handling components of this type. Technicians actually go so far as to place metal straps on their wrists, connected to wires that end in a substantial electrical ground, ensuring that their bodies don't acquire a charge sufficient to "fry" sensitive MOS or CMOS components.

For Nerds Only

Many of the elements found in semiconductors can be mined from the earth. Others are "grown" as crystals under laboratory conditions, a process called *epitaxy*.

Vacuum Tubes

Before the 1960s, *vacuum tubes* prevailed in nearly every electronic device you could find. Even in radio receivers and portable television sets, all of the amplifiers, oscillators, power supplies, and other circuits required tubes (called *valves* in England). A typical vacuum tube ranged from the size of your thumb to the size of your fist. An elaborate vacuum-tube radio was a major appliance, and some of them were as big and heavy as the dresser in your bedroom!

Vacuum tubes are still used in some radio-transmitter and audio power amplifiers, microwave oscillators, and video display units. Tubes work better in certain ways than semiconductor devices do, even today. Tubes can tolerate momentary voltage and current surges and transients better than semiconductors can do. Some popular music bands claim that amplifiers built with vacuum tubes produce richer, truer sound than amplifiers built with semiconductor devices. But tubes have two big drawbacks: They need high voltages to operate, and they consume a lot of power for the actual work that they do.

Did You Know?

Back in the days when vacuum tubes prevailed in electronics, it took from 50 to a few hundred volts of DC to make a vacuum tube function, even in relatively small consumer devices, such as clock radios and portable television sets. These voltages mandated the use of bulky power supplies, and created an electrical shock hazard.

A vacuum tube accelerates electrons to high speed, resulting in electric current. This current can be made more or less intense, or focused into a beam and guided in a certain direction. The intensity and/or beam direction can be adjusted with extreme rapidity, making possible a variety of different useful effects. In any vacuum tube, the charge carriers are *free electrons*, meaning that they don't "orbit" any particular atomic nucleus. Instead, they hurtle like submicroscopic bullets through the tube's internal vacuum.

Before the start of the twentieth century, scientists knew that electrons could carry current through a vacuum. They also knew that hot electrodes would emit electrons more readily than cool ones would. These phenomena went to practical use in the first electron tubes, known as *diode tubes*, for the purpose of rectification.

In any tube, the electron-emitting electrode is called the *cathode*. A wire filament that carries AC, similar to the glowing element in an incandescent bulb, heats the cathode. In some tubes, the filament also serves as the cathode. This type of electrode is called a *directly heated cathode*. In other tubes, the cathode is separate from, and surrounds, the filament. This arrangement is called an *indirectly heated cathode*. The electron-collecting electrode is known as the *anode* or *plate*. The cathode, and by extension the negative DC output of the power supply, is usually connected to a metal chassis that serves to support all the electronic components in the device. The chassis is connected to electrical ground.

Did You Know?

In either the directly heated or indirectly heated cathode type of vacuum tube, electrons get driven off the element by the heat of the filament. Because the electron emission depends on the filament or "heater," tubes need a little while to "warm up." This waiting period is (ironically) about as long as the boot-up time for a typical computer.

A tube's anode comprises a metal cylinder concentric with the cathode and filament. The plate goes to the positive DC power supply terminal, usually through a coil or resistor. The output signal is usually taken from the plate.

Tubes operate at voltages ranging from about 50 volts to several thousand volts. Because the plate readily attracts electrons but is not a good emitter of them, and because the opposite state of affairs prevails for the cathode, a diode tube works well as an AC rectifier or AC-to-DC converter. Although semiconductor diodes have replaced tubes for rectification in most applications, tubes are still used in power supplies that must deliver extreme voltages.

Voltages imposed deliberately on an electrode between the cathode and the plate can control the flow of current in a vacuum tube. This electrode, called the *control grid*, comprises a wire mesh that lets electrons pass through. The control grid interferes with the electrons if it is provided with a voltage that's negative with respect to the cathode voltage. The greater this so-called *negative grid bias*, the more the grid impedes the flow of electrons, and the less current flows in the plate. In most tube-type amplifiers, the control grid receives the AC input signal.

Some vacuum tubes have more than one grid. The extra grids help amplifiers to boost the signals more (in other words, they allow for more *gain*). They also help the amplifier to operate in a more stable manner than a single-grid tube does. A *screen grid* can be added between the control grid and the plate. This grid carries a positive DC voltage, roughly 1/3 that of the plate voltage. The screen grid reduces the tendency of the amplifier to *oscillate* (generate a signal of its own because of feedback inside the tube). The screen grid can also serve as a second control grid, allowing for the injection of two different signals into a tube so that they can be *mixed*, producing new signals at different frequencies or with different characteristics than the originals.

Tube performance can sometimes be improved even more by placing a third grid, called the *suppressor grid*, between the screen grid and the plate. The suppressor grid carries a negative charge with respect to the screen and the plate; usually it's the same voltage as the cathode. The suppressor grid reduces the tendency of a tube to oscillate more than a screen grid alone can do. In addition, a suppressor grid "recovers" stray electrons that "bounce off" the plate upon impact, so that the tube can amplify better than it would without the suppressor grid.

Did You Know?

Old-time radio and TV receivers featured tubes with four or five grids in some circuits. The usual function of such tubes was signal mixing. You'll probably never hear about these devices in modern electronics because solid-state components do all the signal mixing nowadays.

The most common contemporary application of vacuum tubes is in massive amplifiers designed to deliver a great deal of signal output power, especially at high frequencies (in radio and television transmitters, for example) and in big audio systems (for popular music bands, for example).

Vacuum tubes prevail in antique TV receivers and old computer monitors. These devices, which can be large and heavy, are called *cathode-ray tubes* (CRTs). In a CRT, a device called an *electron gun* emits a high-intensity stream of electrons, something like a "flashlight" that "shines" a beam of subatomic particles. This beam gets focused and accelerated as it passes through the holes in donut-shaped anodes that carry high-positive DC voltages. The anodes of a CRT work differently than the anodes of a conventional vacuum tube do. Rather than hitting the anodes, the electrons go right on through, gaining speed with each pass, until they strike a screen with a phosphorescent inner coating. The phosphor glows visibly as seen from the face of the CRT. Internal coils or electrodes carry signals that deflect the electron beam back-and-forth and up-and-down in an intricate pattern at speeds faster than the eye can follow, creating a motion-picture image on the phosphorescent screen.

Some electronics hobbyists enjoy working with antique radios. Certain people like to have a radio broadcast receiver that takes up as much space and weighs as much as a small refrigerator. The big old "boat anchor," sitting on the living room floor, brings back memories of a time when drama shows came over local radio stations. Users had to employ their imaginations as the plots unfolded in the voices and music; video was nothing more than a few roguish inventors' dreams-yet-to-come-true.

Did You Know?

In the days of radio before semiconductor devices took over most of the roles of vacuum tubes, the high internal voltages (required for the tubes to work) caused dust and airborne oil droplets to accumulate on a radio's circuit components as a result of *electrostatic precipitation*. Over the years, the grit, grease, and grime would grow into a gray, fuzzy wax. You might even find a few dead and well-cooked insects in there! Sometimes these insects would cause a circuit malfunction by shorting out something. According to one legend, the term *bug* (in regards to flaws in all sorts of devices and systems, including computer software) arose from this phenomenon.

For Nerds Only

Perhaps you'd like to collect and operate antique radios, just as some people collect and drive vintage cars. However, if you get interested in that stuff, you should know that replacement vacuum tubes are hard to find, and they can also be costly. When your old toy breaks, you'll need to become a spare-parts sleuth.

Digital Logic

Digital logic, also called simply *logic*, is a form of "reasoning" used by electronic machines, particularly devices and systems controlled by computer chips. The most common form of digital logic takes the form of *Boolean algebra*, which uses only the numbers 0 and 1 along with operations called AND (multiplication), OR (addition), and NOT (negation). This system gets its name from the nineteenth-century British mathematician *George Boole*, who supposedly invented it.

- The AND operation, also called *logical conjunction*, operates on two or more quantities. Let's denote it by using an asterisk, for example X * Y.
- The NOT operation, also called *logical inversion* or *logical negation*, operates on a single quantity. Let's denote it by using a minus sign (–), for example –X.
- The OR operation, also called *logical disjunction*, operates on two or more quantities. Let's denote it by using a plus sign (+), for example X + Y.

TABLE **1-1**   Basic Operations in Boolean Algebra*

X	Y	−X	X * Y	X + Y
0	0	1	0	0
0	1	1	0	1
1	0	0	0	1
1	1	0	1	1

*The minus sign represents "NOT"; the asterisk represents "AND"; the plus sign represents "OR." X and Y stand for logical statements (variables).

Table 1-1 breaks down all the possible input and output values for the above-described Boolean operations, where 0 stands for "falsity" and 1 stands for "truth."

In mathematics and philosophy courses involving logic, you'll sometimes see other symbols used for these operations. Conjunction might be denoted with a times sign (×) or a wedge (∨). Negation might be denoted with a "lazy backwards L" (¬). Disjunction might be denoted with an inverted wedge (∧).

All digital electronic devices do Boolean algebra "automatically." These switches, called *logic gates*, can have from one to several inputs and a single output. Table 1-2 summarizes the functions of common logic gates, assuming a single input for the NOT gate and two inputs for the others. Figure 1-15 shows the symbols that engineers use to represent logic gates in circuit diagrams.

TABLE **1-2**   Logic Gates and Their Characteristics*

Gate type	Number of inputs	Remarks
NOT	One	Changes state of input.
OR	Two or more	Output high if any inputs are high. Output low if all inputs are low.
AND	Two or more	Output low if any inputs are low. Output high if all inputs are high.
NOR	Two or more	Output low if any inputs are high. Output high if all inputs are low.
NAND	Two or more	Output high if any inputs are low. Output low if all inputs are high.
XOR	Two	Output high if inputs differ. Output low if inputs are the same.

*In most situations, the logical high state represents "truth," and the logical low state represents "falsity." In electronic terms, high represents the digit 1 and low represents the digit 0.

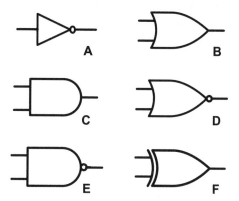

FIGURE 1-14   Symbols for logic gates. At A, a logical inverter or NOT gate. At B, an OR gate. At C, an AND gate. At D, a NOT-OR (NOR) gate. At E, a NOT-AND (NAND) gate. At F, an exclusive-OR (XOR) gate.

For Nerds Only

In so-called *positive logic,* a circuit represents the binary digit 1 with an *electrical potential* of approximately +5 volts DC (called the *high state*), while the binary digit 0 appears as little or no DC voltage (called the *low state*). Some circuits employ *negative logic,* in which little or no DC voltage (low) represents logic 1, while +5 V DC (high) represents logic 0. In another form of negative logic, the digit 1 appears as a negative voltage (such as −5 volts DC, constituting the low state) and the digit 0 appears as little or no DC voltage (the high state, because it has the more positive voltage).

CHAPTER **2**

Tools and Tests

In this chapter, you'll assemble a workbench and learn how to make a few electrical measurements. A multimeter, a soldering iron or gun, a transient suppressor, and some common tools will give you a good start. I'll end this chapter by showing you a trick that can extend the lives of old-fashioned incandescent bulbs.

Your Workbench

Every experimenter needs a good workbench. Mine comprises a piece of plywood 48 inches wide by 24 inches deep and 3/4 of an inch thick, weighted down over the keyboard of an old upright piano in the cellar, and reinforced by chains that go up to the ceiling. I sit at a bar stool while conducting business at the site because the work surface lies almost four feet above the floor.

Your test bench doesn't have to be as strange as mine, of course, and you can put it anywhere you want (within reason), as long as it won't shake or cause you to worry about possible collapse. Stability is of prime importance here; a "card table" will *not* do the job. The surface should consist of an electrically nonconductive material, such as wood or hard plastic, protected in the immediate work area by a tough, fire-resistant place mat or doormat. A desk lamp, preferably the "high-intensity" type with an adjustable arm, completes the ensemble.

Buy a good pair of safety glasses at your local hardware store. Wear the glasses at all times while testing or troubleshooting, or when doing any other experiment or project described in this book. Get into the habit of wearing the safety glasses whether you think you need them or not. You never know when a little piece of wire will fly at one of your eyes when you snip it off with a diagonal cutters! On several occasions, my safety glasses have saved me from trips to the emergency room.

Table 2-1 lists the items that you'll need for a simple, but solid, home electronics workbench. You can find most of these components at hardware and department stores. In a few cases, you might have to visit a Radio Shack store or order a component from their website at www.radioshack.com.

TABLE 2-1   Home Electronics Workbench Items.*

Quantity	Source	Description
1	Department store or hardware store	Pair of safety glasses that seal all the way around your eyes
1	Department store or hardware store	Basic tool kit (hammer, screwdrivers, wrenches, etc.)
1	Department store	Pair of rubber-soled shoes
1	Department store	12-inch plastic or wooden ruler
1	Department store or hardware store	36-inch wooden measuring stick (also called a "yardstick")
1	Department store or hardware store	Large magnifying glass with a good handle
1	Hardware store or Radio Shack	Multimeter with digital numeric readout
1	Hardware store or Radio Shack	Multimeter with analog readout
1	Hardware store or Radio Shack	Soldering gun rated at 100 to 140 watts
1	Hardware store or Radio Shack	Roll of solder for electronic circuits
3	Department store or hardware store	Large roll of electrical tape
1	Department store or hardware store	Diagonal wire cutter/stripper
1	Department store or hardware store	Small needle-nose pliers
1	Radio Shack Part No. 278-1156	Package of insulated test/jumper leads
1	Department store or hardware store	Multi-outlet cord with transient suppressor
1	Department store or hardware store	Pair of thick rubber gloves
5	Department store or hardware store	Steel wool pad without soap
5	Department store or hardware store	Emery paper or emery cloth
1	Department store or hardware store	6-volt lantern battery
1	Department store or hardware store	6-volt lantern bulb

TABLE 2-1   Home Electronics Workbench Items.* (*continued*)

Quantity	Source	Description
1	Department store or hardware store	100-foot, three-wire extension cord
1	Department store or hardware store	6-foot, two-wire extension cord with one plug and three outlets
1	Department store or hardware store	Multiple-outlet power strip with fuse but no surge suppressor
1	Department store or hardware store	25-watt 117-volt incandescent bulb
1	Department store or your house	Common table lamp with switch
1	Hardware store	Two-prong standard-base utility light bulb socket
1	Hardware store	"25-watt-equivalent" 117-volt LED spot light bulb
1	Radio Shack Part No. 276-1104	Package of two rectifier diodes rated at 1 ampere and 600 peak inverse volts

*You can find them at retailers near most locations in the United States.

Warning!   Do not forget to obtain—and wear—a pair of rubber-soled shoes and a couple of rubber gloves. The shoes will ensure that electric current can never flow through your feet to the earth or a concrete floor. Even a seemingly dry concrete floor can conduct enough electricity to present a hazard, especially at utility voltages. The gloves will keep you from getting a shock if you accidentally touch a live wire or surface. (You shouldn't make mistakes like that, but sooner or later, even the best electrician or engineer does.) If you think that I'm acting silly by recommending all of these precautions, that's fine. Go ahead and laugh. Follow these guidelines anyway, and maybe you'll stay alive long enough to laugh at me a lot more in the years to come.

Wire Splicing

The simplest way to splice two single-conductor wires of the same diameter involves stripping the insulation off the ends for about an inch (2.5 centimeters) to expose the bare wire, bringing the bare-wire ends close together and parallel, and then wrapping them over each other to make a *twist splice* as shown in Fig. 2-1. If the wires differ from each other in diameter, you can wrap the smaller wire around the larger one (Fig. 2-2). In any case, once you've made the splice, wrap plenty of electrical tape over the connection to protect it from the elements, to keep the splice from slipping apart, and to make sure that the bare wire doesn't come into contact with anything and cause a short circuit or create a shock hazard.

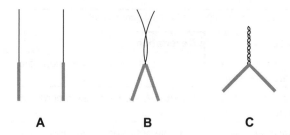

FIGURE 2-1 Twist splice for wires of the same diameter. Bring the exposed wire ends together and parallel (A), loop them around each other (B), and then wrap them over each other several times (C).

When a splice must have some mechanical strength so that the resulting wire can endure "pulling tension," strip about 3 inches (7.5 centimeters) of insulation from the wire ends. Then bring the exposed wires together and parallel from opposite directions, overlapping by about 2 inches (5 centimeters). Loop the wires over each other and then twist the exposed ends around each other several times on either side of the loop center, as shown in Fig. 2-3. Technicians call this type of connection a *Western Union splice*. As with the twist splice, electrical tape can protect the connection from the elements, and also protect you from the risk of electrical shock.

Quick Question, Quick Answer

- What type of splice works best in long spans of wire that put a fair amount of "pull-apart" mechanical stress on the connection?
- You should try to avoid using splices in wire spans. If you have to make a splice that will endure stress, the Western Union method works better than the twist method. Tightly wrap each exposed end around its "mate" about a dozen times. Then *solder* the entire connection and wrap it with electrical tape.

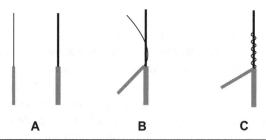

FIGURE 2-2 Twist splice for wires of differing diameter. Bring the exposed wire ends together and parallel (A), loop the smaller wire over the larger one (B), and finally wrap the smaller one several times around the larger one (C).

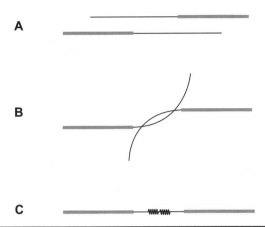

Figure 2-3   Western Union splice. Bring the exposed wire ends close together and parallel from opposite directions (A), loop them around each other (B), and finally wrap them over one another on either side of the center (C).

You should always apply solder to splices if you need a solid, long-lasting electrical and mechanical bond. The soldering process provides extra tensile strength to Western Union splices, and keeps corrosion from degrading the electrical connection as time passes. For large-diameter wires, you should coat both exposed wire ends with solder before making the splice to optimize the electrical bond, a process called *tinning*.

Before you start to solder a wire splice, give your *soldering gun* or *soldering iron* plenty of time to heat up. Soldering guns usually heat up in a minute or less, but soldering irons take several minutes. Once the instrument has reached its working temperature, press its tip against each twist in the splice, allowing the wire to get so hot that the solder flows freely in between the turns as you hold the solder against the wire. Use only enough solder to completely coat or soak the connection. Don't use so much solder that it drips from the connection. Always hold the solder against the wire near, but not exactly on, the tip of the iron or gun.

If you don't apply enough heat to the wire when you apply solder to a splice, you'll end up with a *cold solder joint*. A properly soldered connection has a shiny, clean appearance. A cold joint looks dull or rough. A good solder joint is mechanically solid, while a cold one will likely break apart under stress or repeated flexing.

For Nerds Only

Many electrical equipment failures occur because of cold solder joints, which can exhibit high resistance and/or intermittent conduction. If you find a cold solder joint in an existing splice or circuit, remove as much of the solder as possible, using a heavy wire braid called *solder wick* made especially for this purpose. Then clean the exposed wire surfaces and *resolder* the connection.

Did You Know?

You'll have to use special *aluminum solder* for aluminum wire because aluminum won't adhere to the conventional *soft solder* intended for copper wire. If you need an especially strong mechanical bond, you can use *silver solder*.

Cords and Cables

You'll encounter various types of cable for the transmission of electrical power or signals over short to moderate spans. Let's look at the most common cable configurations.

Lamp Cord

The simplest electrical cable, other than a single wire, is two-conductor *lamp cord*, also known as *zip cord*. It works well with common appliances at low to moderate current levels. Two wires are embedded in rubber or plastic insulation that serves as the *jacket* (Fig. 2-4A). The individual conductors are stranded to help them resist

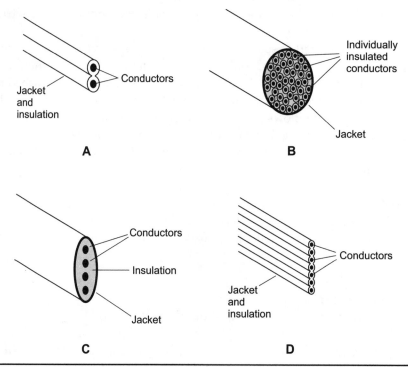

FIGURE 2-4   Two-wire lamp cord (A), bundled multiconductor cable (B), flat multiconductor cable (C), and multiconductor ribbon (D).

breakage from repeated flexing. Some household appliance cords have three conductors rather than two. The third wire facilitates electrical grounding.

Multiconductor Cable

When a cable has several wires, they can be individually insulated, bundled together, and enclosed in a tough external jacket, as shown in Fig. 2-4B. If the cable must have good flexibility, each wire will be stranded. Some cables of this type have dozens of conductors. If only a few conductors exist, they can run parallel to each other and be protected by a tough plastic outer jacket, as shown in Fig. 2-4C, a configuration called *flat cable*. Sometimes, several conductors are molded into a flexible, hard, thin plastic strip, as shown in Fig. 2-4D, an arrangement known as *multiconductor ribbon*. You'll find this type of cable inside high-tech electronic devices, particularly computers. It's physically sturdy, takes up a minimum of space, and efficiently radiates heat away.

Cable Shielding

All of the above-described cable types are *unshielded*. That's a technical way of saying that they can "leak" signals, and they can also pick up signals that come from the outside. For DC and utility AC, unshielded cables usually work okay. But radio, video, and high-speed data generate *electromagnetic (EM) fields* that can "cross over" among the conductors within a single cable, and also between a cable and the surrounding environment. In these situations, you'll want to use *shielded cable* to prevent these troublesome phenomena from taking place.

Cable manufacturers shield an individual wire conductor by surrounding it with a conductive cylinder or conduit made of solid metal, metallic braid (usually copper), or metal foil. A layer of insulating, flexible material such as polyethylene keeps the shield electrically and physically separated from the *center conductor*. Because the conduit and the center conductor run along a common axis, this type of cable has become known as *coaxial cable*. You'll find it in home television and Internet systems all over the United States.

Did You Know?

In some multiconductor cables, a single shield surrounds all the wires. In other cables, each wire has its own dedicated shield. The entire cable can be surrounded by a hose-like copper braid in addition to individual shielding of the wires. *Double-shielded cable* is surrounded by two concentric hose-like braids separated by insulation.

Coaxial Cable

Coaxial cable, also called *coax* (pronounced "*co*-ax"), serves as the transmission medium of choice for *radio-frequency* (RF) signal transmission because of its excellent

EM shielding characteristics. You can consider any AC whose frequency exceeds a few kilohertz (thousands of hertz) to lie within the so-called *RF spectrum*.

You'll find coax in *community-antenna television* (CATV) networks and in some computer *local area networks* (LANs). Amateur, marine, Citizens Band, and government radio operators use coax to connect their transceivers, transmitters, and receivers to external antennas. Lightweight coax works well in high-fidelity sound systems when you want to interconnect components, such as amplifiers, microphones, tuners, and speakers, because it keeps the audio signals in your system while keeping stray electrical and RF noise out.

In a typical coaxial cable, a cylindrical shield surrounds a single center conductor. In some cable types, solid or foamed polyethylene insulation, called the *dielectric*, keeps the inner conductor running exactly down the center of the cable (Fig. 2-5A). Other cables have only a thin tubular layer of solid polyethylene just inside the shield (Fig. 2-5B), so that air makes up most of the dielectric medium.

The most sophisticated coaxial cables have a solid metal shield that resembles a pipe or conduit. Communications engineers call it *hardline*. It's available in larger diameters than typical coaxial cables, and transmits signals with excellent efficiency. You'll find hardline in high-power, fixed radio and television transmitting installations.

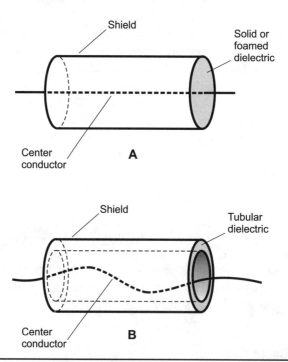

FIGURE 2-5   At A, coaxial cable with solid or foamed insulation (called the dielectric) between the center conductor and the shield. At B, similar cable in which air makes up most of the dielectric medium.

Quick Question, Quick Answer

- Can coaxial cable effectively carry DC?
- Yes. In some situations, coaxial cable works better as a DC-carrying medium than a single wire or lamp cord does. But coax costs more money, and it's harder to work with mechanically.

Did You Know?

When you interconnect sensitive electronic devices with lengths of cable, you'll sometimes need to keep EM fields away from the currents flowing in the cable. If the cable conductors are exposed to strong EM fields, the fields can induce additional (and unwanted) current, causing certain types of devices to malfunction. An example is the cable used to connect the speakers of a hi-fi sound system to the amplifier. The use of coaxial cable in place of typical speaker cable (which resembles lamp cord) can eliminate, or at least reduce, the risk that a strong EM field will induce stray AC into the amplifier by means of the speaker wires.

Fact or Myth?

Some people will tell you that coax shielding helps to keep harmful EM fields from leaking out, adversely affecting your body organs and contributing to chronic diseases and complaints. If you've heard any variant of this tale (and a lot of them exist), you can take it seriously, but only up to a point. Yes, shielding will indeed reduce your exposure to EM fields from electronic circuits. However, that fact doesn't necessarily imply that you must rewire your home's entire utility system with coaxial cables if you want keep the EM fields from slowly killing you (but in theory, it wouldn't harm you except maybe in terms of your bank account balance!). Scientists and medical professionals disagree about the extent to which EM field exposure has contributed to chronic maladies, such as migraines, lassitude, depression, and even cancer, in recent decades. To use a time-worn phrase, "The jury remains out." In other words, I can't tell you for sure whether any of these effects are real, or not.

Cable Splicing

When you want to splice cord, multiconductor cable, or ribbon cable, you can use a Western Union splice for each individual conductor. You should wrap all splices individually with electrical tape, and then wrap the entire combination with more electrical tape afterwards. Insulating the individual connections ensures that no two conductors will come into electrical contact with each other inside the splice.

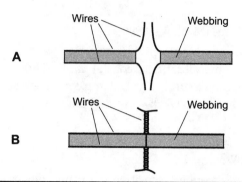

FIGURE 2-6 Twist splice for two-wire cord or cable. Bring the ends together (A) and twist them at right angles (B). Then solder, trim, fold back, and insulate the whole connection.

Insulating the whole combination keeps the cable from shorting out to external metallic objects and provides additional protection against corrosion and oxidation.

You can twist-splice two-conductor lamp cord or ribbon cable. First, bring the ends of the cables together and pull the conductors outward from the center, as shown in Fig. 2-6A. Then twist the corresponding conductor pairs several times around each other, as shown in Fig. 2-6B. Solder both connections and then trim the splices to lengths of approximately 1/2 inch (1.2 centimeters). Insulate each twist splice carefully with electrical tape. Then fold the twists back parallel to the cable axis in opposite directions. Finally wrap the entire connection with electrical tape to insulate it and hold the splices in place.

In order to minimize the risk of a short circuit occurring between conductors in a multiconductor cable, you can make the splices for each conductor at slightly different points along the cable. But in any case, you *must* ensure that each conductor at the end of the first cable connects to the corresponding conductor at the end of the second cable! This process can get confusing, especially if the color-coding scheme for the wires in the first cable differs from the color-coding scheme for the wires in the second cable. Always double-check your work before insulating the connection. It's a huge hassle to take apart a complicated cable splice.

Plugs and Connectors

We're all familiar with the *plugs* attached to the ends of appliance cords. These connectors have two or three prongs that fit into *outlets* having receptacles in the same configuration. The plugs are called *male connectors*, and the receptacles into which the plugs fit are called *female connectors*.

Appliance plugs and outlets provide temporary electrical connections. They can't offer reliability in long-term applications because the prongs on any plug, and the metallic contacts inside any receptacle, will eventually *oxidize* or *corrode*. A new male plug has shiny metal blades, but an old one has visibly tarnished blades. Similar oxidation occurs in the "slots" of a female receptacle, even though you can't see it.

A *clip lead* (pronounced "leed," which means "wire") comprises a short length of flexible wire, equipped at one or both ends with a simple, temporary connector. Clip leads do a poor job in permanent installations, especially when used outdoors because corrosion occurs easily, and the connector can slip out of position. In addition, clip-lead connections can't carry much current. Clip leads are used primarily in DC and low-frequency AC applications at low voltage and current levels.

A *banana connector* is a convenient single-lead connector that slips easily in and out of its receptacle. The "business part" of a *banana plug* (male connector) looks something like a banana (or, if you prefer, a cucumber), as shown in Fig. 2-7. It's like a "round peg." *Banana jacks* (female receptacles, not shown in the figure) have a cylindrical shape; they're "round holes." You'll sometimes find banana jacks inside the *screw terminals* of low-voltage DC power supplies.

Warning! Banana connectors, like clip leads, are intended for low-voltage, low-current, and short-term indoor use only. Never use these connectors with high voltages. If you do, the exposed conductors will pose an electrical shock hazard.

A *hermaphroditic connector* is an electrical plug/jack with two or more contacts, some of them male and some of them female. Usually, hermaphroditic connectors at opposite ends of a single length of cable look identical when viewed "face-on." However, the pins and holes have a special geometry, so you can join the two connectors in the correct way only. This "can't-go-wrong" feature makes hermaphroditic connectors ideal for polarized circuits, such as DC power supplies, and for multiconductor electrical control cables.

Phone plugs and jacks find extensive use in DC and low-frequency AC systems at low voltages and low-current levels. In its conventional form, the male phone plug (Fig. 2-8A) has a rod-shaped metal *sleeve* that serves as one contact, and a spear-shaped metal *tip* that serves as the other contact. A ring of hard-plastic insulation separates the sleeve and the tip. Typical diameters are 1/8 inch

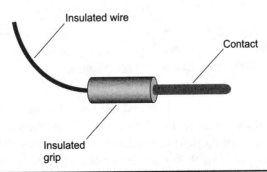

Insulated wire

Contact

Insulated grip

FIGURE 2-7 Banana connectors work well in low-voltage DC applications. The single contact (the plug, shown here) slides into a cylindrical receptacle (the jack, not shown here).

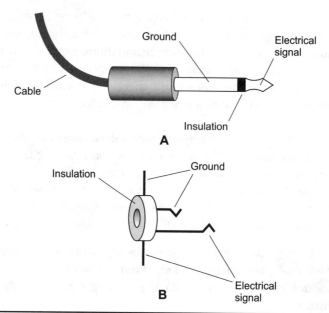

FIGURE 2-8   At A, a two-conductor phone plug. At B, a two-conductor phone jack.

(3.175 millimeters) and 1/4 inch (6.35 millimeters). The female phone jack (Fig. 2-8B) has contacts that mate securely with the male plug contacts. The female contacts have built-in spring action that holds the male connector in place after insertion.

Engineers originally designed the phone plug and jack for use with two-conductor cables. In recent decades, three-conductor phone plugs and jacks have become common as well. They're used in high-fidelity stereo sound systems and in the audio circuits of multimedia computers and radio receivers. The male plug has a sleeve broken into two parts along with a tip, and the female connector has an extra contact that touches the second sleeve when you insert the plug into the jack.

Did You Know?

The term *phone* comes from the original application of phone connectors: centralized community *telephone switchboards* that were manipulated by human operators prior to the advent of direct dialing.

Phono plugs and *jacks* are designed for ease of connection and disconnection of coaxial cable at low voltages and low-current levels. You can simply push the plug onto the jack, or pull it off. These connectors work in the same situations as phone plugs and jacks do, but they offer better shielding for coaxial-cable connections. Phono plugs and jacks are also known as *RCA connectors*, named after their original designer, the *Radio Corporation of America* (RCA).

Did You Know?

The term *phono* comes from the earliest design application of phono plugs and jacks: interconnecting the audio components of *phonographs* ("record players") in the early part of the twentieth century for music recording and reproduction.

If a cable contains more than three or four conductors, you can put a *D-shell connector* on either end. These connectors are available in various sizes, depending on the number of wires in the cable. The so-called *ports* on a personal computer, especially the one intended for connecting the *central processing unit* (CPU) to an external *video display*, commonly employ D-shell connectors, which have the characteristic appearance shown in Fig. 2-9. The trapezoidal shell forces you to insert the plug correctly; you can't accidentally put it in upside-down. The female socket has holes into which the pins of the male plug slide. Screws or clips secure the plug once it's in place.

For Nerds Only

As previously mentioned, plug contacts gradually oxidize (combine with oxygen in the air), especially if exposed to the outdoors or used in a humid indoor environment. You can tell when oxidation has taken place because the metal prongs appear dark or discolored. You can remove the oxidation, which forms in a thin layer, by rubbing the contacts with fine-grain sandpaper, emery paper, or steel wool, and then wiping the contacts off with a dry cloth. You should continue the "sanding" process until bright metal shows everywhere on the exposed parts of the contacts. If the contact prongs are too small or too closely spaced for "sanding," you can use specially formulated contact cleaner to remove the oxidation layer. This type of cleaner can be purchased in a good hardware or electronics store, such as Radio Shack. Sometimes, a solution of table salt and vinegar will do the job, but you must be sure to rinse all of that salt off the connectors after you use a solution like that. Otherwise, the chlorine in the salt will accelerate the corrosion process.

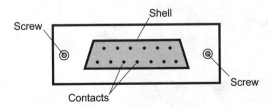

FIGURE 2-9   You can recognize a D-shell connector by its characteristic trapezoidal shape. The number of contacts can vary.

Meet the Multimeter

You can buy a multimeter at a well-stocked hardware store, at most Radio Shack retail outlets, and through various websites. A typical multimeter is about the same size as a pocket calculator or small e-book reader. *Analog* meters have old-fashioned "needle-and-scale" type readouts. *Digital* meters actually show you the numbers. The type that you buy depends on your personal preference. I have both an analog multimeter (at left in Fig. 2-10) and a digital multimeter (at right in Fig. 2-10).

Warning!   Whenever you use a multimeter to test any circuit in which you suspect that the voltage will exceed 12 V (the level produced by an automotive battery), wear your rubber gloves and a good pair of rubber-soled shoes. That way, you'll make absolutely sure that you can't receive a dangerous shock. You can never tell when high voltage will "sneak in" somewhere and try to zap you. (It's almost as if gremlins lurk out there in the electrical cosmos, waiting for an opportunity to clobber you the instant you let your guard down.) In addition, by insulating yourself completely from the circuit under test, you'll ensure that your body's *internal resistance* can't throw off the meter reading. This phenomenon can occur especially when you measure extremely high resistance values or low-current levels.

My analog multimeter has several graduated scales, and a rotary switch that offers 14 settings. When the switch points straight up, the meter is turned off. (You

FIGURE 2-10   At left, an analog multimeter. At right, a digital multimeter.

should keep the meter turned off when you're not using it because it contains a battery that will gradually discharge if you leave the meter powered-up, even if it sits idly on a shelf.) Going clockwise from the "OFF" position, the switch allows measurement of the following thirteen quantities, in order:

1. DC volts (DCV) from 0 to 10
2. DCV from 0 to 50
3. DCV from 0 to 250
4. DCV from 0 to 500
5. AC volts (ACV) from 0 to 500
6. ACV from 0 to 250
7. ACV from 0 to 50
8. DC milliamperes (DCmA) from 0 to 250, where one milliampere (1 mA) equals a thousandth of an ampere (0.001 A)
9. DCmA from 0 to 25
10. Battery test (BAT) for 1.5-volt (1.5-V) cells
11. Battery test (BAT) for 9-volt (9-V) batteries
12. Resistance in thousands of ohms (×1 k)
13. Resistance in tens of ohms (×10)

Quick Question, Quick Answer

- When I look at an analog multimeter, I notice that the resistance scale goes backwards. That is, 0 is on the far right-hand end of the scale, and the left-hand end has a little sideways 8. Why does the resistance scale go backwards?
- When you measure resistance, the meter reading depends on the amount of current flowing through the device that you're testing. The meter's internal battery forces a certain current through the component, and that current depends on the resistance. As the resistance goes down, the current goes up. That's why the scale goes backwards. The sideways 8 at the left-hand end of the scale means "infinity ohms." That's an open circuit, where no current flows at all.

On any analog meter that can measure resistance, you'll find a little knob that allows you to adjust the meter for the correct zero reading. It will say "0 ADJ" or something like that. The "horseshoe" symbol is an uppercase Greek letter omega, which stands for "ohms." To use this control, set the meter switch to the resistance range that you intend to use, short the red and black test probes directly together, and turn the knob until the meter needle goes all the way to the right-hand end of the resistance scale and comes to rest at the hash mark for 0 ohms. Don't turn the knob past that point. The needle should hover exactly over the 0 marker. Use this adjustment control whenever you change the meter from one resistance scale to another, and also if you haven't used your meter for awhile. Make it a habit, like adjusting the rear-view mirrors in a car before you drive it. As the battery grows weak, the meter's accuracy will degrade unless you take advantage of this control before every resistance test.

Did You Know?

When you want to use any meter to measure a quantity of known type (DC volts, for example), you should start with the highest scale and work your way down. That way, if the quantity that you want to measure exceeds the maximum scale value, the meter's needle won't forcibly hit the pin at the top end.

Caution! When you want to measure the resistance between two points in a circuit, make sure that the device under test is powered down (switched off) before you connect a multimeter to it. Otherwise, you'll probably get an inaccurate reading. You might even damage your meter, disrupt the operation of the circuit under test, or do both.

My digital multimeter has a rotary switch with 20 positions. As with the analog meter, the top switch position represents "OFF." Going clockwise from there, the switch allows for measurement of nineteen quantities, in this order:

1. AC volts (V~) from 0 to 500, with a special insert for the red test-probe wire
2. V~ from 0 to 200
3. DC amperes (A—) from 0 to 200 microamperes (the switch says 200µ), where one microampere (1 µA) equals a millionth of an ampere (0.000001 A)
4. A— from 0 to 2000 µA (the switch says 2000µ)
5. A— from 0 to 20 mA (the switch says 20m)
6. A— from 0 to 200 mA (the switch says 200m)
7. A— from 0 to 10, with a special insert for the red test-probe wire (the switch says 10 A)
8. A blank spot that doesn't do anything as far as I know
9. Diode test (a diode should conduct in one direction but not the other)
10. DC resistance in ohms (Ω) from 0 to 200
11. Ohms (Ω) from 0 to 2000
12. Ohms (Ω) from 0 to 20,000 (the meter says 20k)
13. Ohms (Ω) from 0 to 200,000 (the meter says 200k)
14. Ohms (Ω) from 0 to 2,000,000 (the meter says 2000k)
15. DC voltage (V—) from 0 to 200 millivolts (the switch says 200m), where one millivolt (1 mV) equals a thousandth of a volt (0.001 V)

16.   V— from 0 to 2000 mV (the switch says 2000m)
17.   V— from 0 to 20
18.   V— from 0 to 200
19.   V— from 0 to 600, with a special insert for the red test-probe wire

Did You Know?

The red test-probe wire should always go to the more positive point in a DC circuit or system, whether you measure current or voltage. In an AC situation, it doesn't matter which probe goes where. When you measure resistance, it sometimes matters, and sometimes doesn't.

For Nerds—or Not

If you really want to know how to get the best use out of your multimeter, read the instruction manual that came with it. You'll learn something new every time you look at that manual. You might also avoid a mistake that could cause circuit damage or a destroyed meter.

Continuity Testing

The simplest test that you can do with a multimeter involves finding out whether two different points in an electrical circuit or system are *shorted out* to each other. In this context, "shorted out" means "directly connected," so that current can flow perfectly well from one point to the other in either direction. That state of affairs translates to no (zero) resistance between the two points.

Imagine two points called X and Y. You want to find out if they're directly connected. In order to do a *continuity test* (a test to see whether current can flow perfectly well) between X and Y, perform the following steps in order.

1.   Put on your rubber gloves and shoes.
2.   Set your multimeter to measure the highest AC voltage that it can.
3.   Touch the black test lead's probe tip to point X. (Again, in this context, engineers pronounce the word "lead" as "leed." It means "wire.")
4.   Touch the red test lead's probe tip to point Y, while leaving the black probe on X.
5.   Look at the meter reading. If it shows anything other than 0, you know that an AC voltage exists between points X and Y, so they can't possibly be directly shorted to each other.
6.   If you do see 0 as the result for step 5, switch the meter to the next lower AC voltage function and repeat the test.

7. If you get 0 again, repeat steps 3 through 5 with all the AC voltage functions that the meter has, going down until you've tested at the lowest AC voltage setting. You should always get a reading of 0. If you don't, you know that some AC voltage exists between the two points, so they can't possibly be shorted directly to one another.

8. Assuming that you've seen readings of 0 for all the AC voltage settings, repeat the tests with the DC voltage functions, starting with the highest one and working your way down to the lowest one. Test for DC voltage in both directions: first with the black lead to X and the red lead to Y, and then the other way. You should always see a meter reading of 0.

9. If you ever get any DC voltage besides 0 between the two points, you know that a perfect short circuit does not exist between them. A voltage can never arise between two points that are directly connected to each other.

10. If you do get 0 for all of the AC and DC voltage settings, switch your meter to the highest resistance function.

11. Touch the two test-probe tips to each other. If you're using an analog meter, tweak the "0 ADJ" knob so that the meter indicates 0. If you're using a digital meter, make sure that the display indicates a value of 0.

12. Touch the black test lead's probe tip to point X, and the red test lead's probe tip to point Y. You should get a reading of 0.

13. Reverse the test leads so that the red one goes to X and the black one goes to Y. You should again get a reading of 0.

14. Assuming that you've gotten readings of 0 in both of the previous steps, set the meter to the next lower resistance function and repeat steps 11 through 13. You should again see readings of 0 between the two points in both directions.

15. Repeat steps 11 through 13 for all the rest of the meter's resistance settings, working your way down to the lowest one.

16. If you always get readings of 0, without fail, for every step in this process, then you can have confidence that the two points X and Y are directly "shorted out" to each other.

17. If you ever see any meter reading other than 0 for the condition between points X and Y, you know that the two points *are not* perfectly shorted out. Maybe there's some resistance between them but no voltage; maybe there's some DC voltage between them; maybe there's some AC voltage between them. In any case, the foregoing set of steps will reveal the truth.

Caution!   Never start any series of tests with your meter set for anything besides the highest AC voltage function. After you've checked for AC voltage, then check for DC voltage. Finally, you can check for resistance, assuming that all the AC and DC voltage settings yield meter readings of 0. (If they don't, you'll get false resistance readings, and you might damage your meter.)

Warning! Never start any sequence of tests by setting your meter to measure current (amperes or milliamperes). When you do that, you place a short circuit between the two points under test. That action could disrupt the circuit under test, blow a fuse somewhere, or, at the worst, ruin your meter and cause damage to the circuit under test.

Measuring DC Resistance

Now it's time to measure some resistance values with the *ohmmeter* (resistance measurement) function of your multimeter. Normally, you'll check resistance values for individual components, away from external circuits. You should always isolate components when you want to know their resistance values. If you try to measure the resistance of a component when it's connected in a circuit with other things, you can't be certain that you're testing the resistance of that one component all by itself. You might end up checking the resistance through several different circuit paths at once, and you'll have no way of knowing which components have which resistance values.

You can check the DC resistance between the prongs of the plug in a common household appliance to see whether or not its input circuits make a connection. The simplest example is a table lamp with an incandescent bulb in it. Switch the lamp off, and unplug it from the electrical outlet. Set it up on your workbench. Then follow the same procedure as you would do for the resistance part of the continuity test described above (steps 10 through 17). You needn't test for AC or DC voltages in this situation because a lamp, disconnected from the electric utility, won't carry any voltage!

When you follow steps 10 through 17 in the continuity test sequence with a lamp that's switched off, you should end up with an open circuit. Now, with the ohmmeter still set at its lowest resistance range, turn the lamp switch on, so that the bulb would light up if the lamp were plugged in. You should see some finite, nonzero resistance indication on the meter. That's the resistance of the bulb filament when it's "cold." If you want to see a little more variety in this experiment, try the same procedure with a three-way lamp and one of those triple-brilliance incandescent bulbs (such as 30, 70, and 100 watts). You should observe different resistance values for each of the lamp switch settings.

For Nerds Only

When you are switching your meter among its various resistance-measurement ranges, see how the reading changes when it shows some finite, nonzero value. If you have an analog meter, the needle will move when you change the resistance range, but you'll get the same actual resistance number when you incorporate the switch instructions (×10 versus ×1k, for example). If you have a digital meter, you'll see the same resistance value regardless of the switch position, but the number of displayed digits, and therefore, the precision of your measurement, might change.

Now, try measuring the resistance of some other appliances, but always while they're disconnected from the electrical utility. How about an electric can opener? Or a cell-phone charger? How about your big-screen TV set, or your computer printer? Does it make a difference whether or not you've switched the device "on"? Always make sure to test the device in both the "on" mode and the "off" mode. And make sure that you know exactly what action will turn the thing "on"! My electric can opener, for example, requires that I press a lever down to make the motor go. Otherwise it's "off."

Did You Know?

Most electronic devices, such as radios, TV sets, and computers, intended for plugging into a wall outlet directly, actually run on DC, not AC. Therefore, they contain specialized power supplies. Such a power supply usually has a transformer whose primary winding goes straight to the wall outlet (sometimes through a switch), so when you measure the resistance between the prongs of the plug, you're looking at the DC resistance through the transformer's primary winding. Sensitive devices might have a fuse in the power cord, which is hidden inside the cabinet, and connected in series with the transformer's primary. If such a fuse blows out, the device won't work, and when you test its plug resistance, you'll get "infinity." An open circuit between the prongs of an electronic device's plug, even when the thing is "switched on," might indicate nothing more than a blown internal fuse. However, it could also indicate something more serious, such as a damaged transformer.

Measuring DC Voltage

When you want to use a multimeter to measure a DC voltage that you know is low, such as that from a single flashlight cell, lantern battery, or "transistor" battery, you don't have to bother with rubber gloves and rubber-soled shoes. But you should wear them if you work with any circuit that carries more than 12 V, or that you suspect might carry more than that voltage. If you want to measure the voltage of an automotive battery, you should wear safety glasses because if you make a mistake, you can cause sparks to fly from that battery, and those sparks can ignite hydrogen gas that might seep out of the battery's electrolyte.

As long as you know that the voltage source you're testing produces DC and not AC, you can set your multimeter for its highest DC voltage range to start out with. Make certain that you have set the meter to measure voltage, and not current or resistance! I've accidentally set a multimeter to measure resistance when I wanted to measure voltage, and caused the fuse inside the meter to blow out as a result. Then I had to go downtown and find a replacement fuse before the meter would work again. As things turned out, it was quite a chore to find that doggone fuse. Save yourself that kind of inconvenience. Always double-check your meter settings before you touch its probe tips to anything!

Place the black test lead's probe tip against the terminal that you think is negative, and place the red lead's probe tip against the terminal that you think is positive. Then work your way down the DC voltage ranges until you get a reading that's easy to read (on an analog meter) or that gives you the best accuracy (on a digital meter), as shown in Fig. 2-11A. If you've connected the probes backwards by mistake (black to positive and red to negative), an analog meter's needle will "pin out" to the left (Fig. 2-11B), a meaningless reading. Most digital meters will give you a negative number for the voltage in that sort of situation, but it'll probably be accurate.

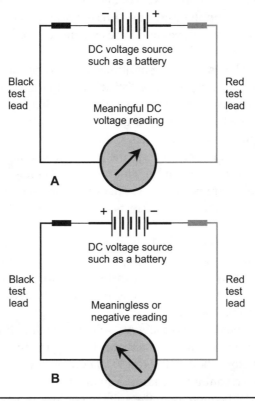

FIGURE 2-11 Correct method of DC voltage measurement (at A) and incorrect method of measurement (at B). Sometimes, you have to find the right result by trial and error!

Did You Know?

In order to make sure that you get an accurate reading for DC voltage (or any other quantity that you measure with a meter), you'll always do best if you see a positive reading that "seems to land in the ball park." You can paraphrase Murphy's law here: If you think that you might have done something wrong during an electrical measurement session, then you probably have! In that case, you had better start the test sequence all over again.

Measuring AC Voltage

When you want to measure AC voltage, you can do it in pretty much the same way as you measure DC voltage. Just make sure that you set the meter for AC volts, not DC volts. If you know that you'll always deal with 12 V or less, you don't have to don any rubber gloves, heavy boots, hazmat suits, or other bulky adornments. It's never a bad idea, though, to double-check the meter function switch to ensure that it's set where you want it to be set.

By far the most common AC voltage measurements are done with household utility circuits and appliances. They present a potentially lethal shock hazard, so you should wear your gloves and shoes whenever you test anything having to do with utility wiring.

Figure 2-12 shows an analog multimeter that's checking the AC voltage at a wall outlet. The meter's test-probe tips "stick" right in the slots! In this case, the probes are in the upper-rightmost of the six outlets. The black lead goes to the longer slot (the one on the left); the red lead goes to the shorter slot (the one on the right). However, as previously mentioned, it doesn't matter which lead goes where when you work with AC. You can reverse the leads and get the same meter reading.

This particular analog multimeter has an AC voltage switch setting where the full-scale reading represents 250 V. That's the one used in the scenario of Fig. 2-12. As you can see, the meter needle rests in the middle of the range, indicating 125 V, give or take a little.

The meter shown in Fig. 2-12 has an AC switch position that registers 500 V at full scale. It will measure 125 V okay with the switch set there, but the needle will only go 1/4 of the way up the scale instead of halfway. As a result, the reading will be less accurate because you'll have to interpolate between more closely spaced marks on the meter's scale.

This same meter also has a low AC voltage switch setting where full scale equals only 50 V. You don't want to use that setting to measure common utility voltage (which ranges from about 110 V to 130 V) because the needle will "hit the pin" and you won't get a meaningful reading. You must always use a full-scale meter setting that's at least a little more than the voltage you want to measure.

Figure 2-12   Measurement of AC voltage. In this example, the meter probe tips are "sticking" into the upper-rightmost outlet of the six outlets.

Warning!   If you intend to measure any voltage (AC or DC) that you think might exceed 12 V, or if you want to check any voltage whose value you don't know "anywhere in the ball park," you absolutely must wear your gloves and shoes. Prepare for the worst possible state of affairs. You're better off overprotecting yourself than getting clobbered by an unexpected electromotive force!

Measuring Direct Current

When you want to measure the current that flows in a circuit, you must insert your multimeter somewhere in the circuit. That means you'll have to break the circuit and connect the meter in series at the point where you've made the break. Figure 2-13 shows a simple example in the form of a schematic diagram, in which a multimeter measures the current in a circuit containing a battery and a light bulb (lamp).

You'll rarely have occasion to measure the current in any electrical circuit. If you do encounter a situation where you must know how much current a certain

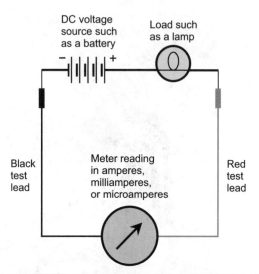

DC voltage
source such
as a battery

Load such
as a lamp

Black
test
lead

Meter reading
in amperes,
milliamperes,
or microamperes

Red
test
lead

Figure 2-13   Measurement of direct current. Always put a load between the meter and the voltage source; never connect the meter directly to the source! Also, make sure you get the polarity correct if you use an analog meter.

circuit carries, you'll always end up checking direct current (DC), such as you get from a battery or a specialized power supply. The measurement of AC requires equipment that's more sophisticated than anything this book deals with; and besides that, beyond the price range of the everyday consumer.

Most battery-powered electronic devices work with 12 V or less, so measuring the current in them presents no shock hazard. You can dispense with the protective gear (gloves and shoes) in these situations for current measurement. But if you ever want to measure the current in a device that has anything to do with a source of more than 12 V, you had better wear all that stupid stuff, even if you think you'll never get exposed to anything more than 12 V. You never know when a cell-phone charger or power supply will develop an internal short circuit that will put raw AC utility voltage onto the low-voltage lines.

Here's a simple experiment that you can perform to demonstrate how a multimeter can measure DC. Obtain a 6-V lantern battery and a 6-V lantern bulb to go with that battery. You can get these items at a hardware store (see Table 2-1). Then go through the following steps, in order.

1. Connect a clip lead (see Table 2-1 again) between the negative battery terminal and the outer, screw-in part of the lantern bulb.
2. Set your multimeter for the highest DC current setting. Make sure that you've selected the current function (amperes or milliamperes). If you select voltage by mistake, the lamp won't light up. If you select one of the resistance functions, you might blow your meter's fuse, and even if you don't, the reading that you get will make no sense.

3. Connect another clip lead between the positive battery terminal and the red test lead's probe tip for your multimeter.
4. Touch the black test lead's probe tip to the contact at the very bottom of the bulb. The bulb should glow at its normal brilliance.
5. Note the meter reading. If you have an analog meter, the needle might not move much, but try to get an idea of the current level anyway.
6. If the current reading appears to be smaller than the next-lower meter current setting, switch the meter to that lower-level function and check the current again.
7. Repeat the preceding step if applicable, until you get the most accurate current reading possible.
8. Don't let an analog meter's needle hit the pin! Always start with the highest current function and work your way down.

For Nerds Only

The ideal voltage-measuring meter would constitute a perfect open circuit, and the ideal current-measuring meter would constitute a perfect short circuit. In other words, the *ideal voltmeter* would have infinite resistance, and the *ideal ammeter* would have zero resistance. In the real world, engineers can design meters that approach these ideals, but they can't quite reach perfection.

Did You Know?

You can easily forget to switch a multimeter off after you've finished doing tests. A couple of unnecessary (and inconvenient) meter battery replacement episodes improved my memory in this respect. You don't have to endure the inconvenience that I did. Switch your meter off when you're done using it!

Testing Fuses

If you intend to check a fuse that forms part of your household utility circuit, or that serves as part of a high-voltage appliance, put on your gloves and shoes before you do anything else! Remove the fuse from its socket and then go through the following steps in order with your multimeter:

1. Switch your meter to the lowest resistance (ohmmeter) function
2. Touch the two test-probe tips to each other. If you have an analog meter, manipulate the "0 ΩADJ" control so that the meter needle floats on the hash mark for 0. If you're using a digital meter, make sure that it gives you a reading of 0.

3.   Touch the black meter lead's tip to one of the fuse contacts, and the red meter lead's tip to the other fuse contact. Hold both probe tips in place against the fuse contacts.

4.   If you get a reading of 0, the fuse is *probably* okay.

5.   If you get a reading of anything other than 0, the fuse is *definitely* bad.

Did You Know?

A good fuse will *always* act like a perfect short circuit when you test it with an ohmmeter. A bad fuse will *usually* appear as an open circuit, but once in awhile you'll encounter a mischievous or defective fuse that tests good but that's actually bad. It's on the verge of destruction even though you can't see it or find out about it with an ohmmeter. It will "blow" the instant you put it in an active circuit. Then you won't have to wonder about it any more!

Incandescent Bulb Saver

A single component no larger than a vitamin pill can greatly prolong the life of an old-fashioned household incandescent bulb. That "miracle worker" is a diode of the same type that you find in power supplies, and which you learned about in Chapter 1. You'll need a 6-foot (2-meter), two-wire extension cord with a single plug on one end and three outlets in the other end; a power strip with a switch, fuse, and several outlets (but *without* a surge suppressor); a common household lamp; a 25-watt (25-W) incandescent bulb; and a rectifier diode rated at one ampere (1 A) and 600 peak inverse volts (600 PIV). You'll also need a diagonal cutter and some electrical tape.

To begin, make sure that the cord and the power strip are not plugged in to a wall outlet or anything else. Cut one of the wires in the extension cord with your diagonal cutter. Be careful to cut only one wire, and don't nick the other wire! Then pull the ends of the broken wire away from the other wire ("unzip" the "zip" cord) for 2 or 3 inches (5 to 7.5 centimeters) on either side of the cut you've made. Strip off 1 inch (2.5 centimeters) of insulation from each of the two cut ends. Install the diode in the side of the cord with the single broken wire, as shown in Fig. 2-14A. Wrap the splices individually with electrical tape so that the wire and diode are completely covered. Then wrap electrical tape around a span of cord that's long enough to secure the diode in place and make the cord almost like new. You now have a *half-wave rectifier* for AC. Figure 2-14B is a schematic diagram of the device.

Plug the rectifier cord into the power strip, and then plug the power strip into a standard utility outlet. Plug a lamp into one of the outlets at the end of the cord. Install a 25-W bulb in the lamp. Switch the power strip and the lamp on. The bulb will glow at somewhat less than normal brilliance. The dimming effect occurs because half of the AC wave gets blocked or "chopped off" by the diode, so current

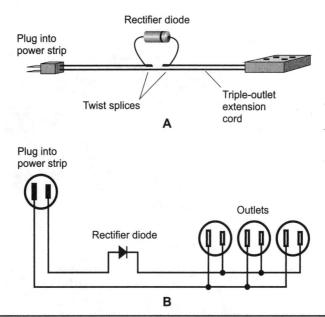

FIGURE 2-14   When you make the rectifier cord as shown at A, you must complete the twist splices and insulate them with electrical tape before you connect the assembly to a source of power. Drawing B shows you the engineering circuit diagram (schematic diagram) of the completed rectifier cord.

flows during only half of each AC cycle. The bulb therefore operates at lower-than-normal *effective voltage*, allowing it to "loaf."

Quick Question, Quick Answer

- It's one thing to reduce the voltage across an incandescent bulb, but quite another thing to actually make it last longer. Does this device really work, or is it nothing more than a gimmick?
- I have used this type of rectifier cord with a desk lamp in my attic office, along with a 15-W bulb to serve as a night light. That bulb stayed aglow continuously, all day and all night, for more than a year! I've never seen an unprotected incandescent bulb last anywhere near that long.

For Nerds Only

As you learned in Chap. 1, the effective voltage produced by a fluctuating or alternating electrical source is technically known as the *root-mean-square*, or RMS, voltage. When you have a sine wave of the sort that you find in household utility circuits, the RMS voltage is approximately 0.707 times the *positive peak*

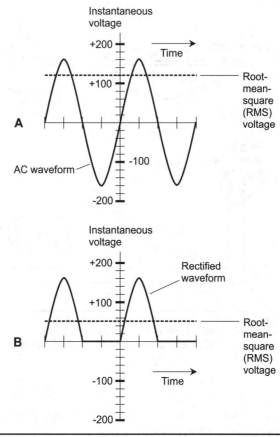

FIGURE 2-15 Here's how half-wave rectification works to reduce the effective voltage of utility electricity. Drawing A shows the AC wave as it appears at the power strip's input. Drawing B shows the "chopped-off wave" as it appears at the cord outlets and across the lamp. Dashed lines indicate the effective (RMS) voltages.

voltage, or about 0.354 times the *peak-to-peak voltage.* When you apply a sine-wave source of electricity to the input of your new rectifier cord, the diode conducts during only half of the cycle. During the other half of the cycle, the diode behaves as an open circuit. Figure 2-15 illustrates the situation. At A, you see a graph of the AC input wave. At B, you see a graph of the rectified output wave. When you burn the incandescent bulb at the lower effective voltage, the filament doesn't get as hot as it would ordinarily, so it lasts longer. Of course, the bulb doesn't burn as brightly, but you should expect that sort of sacrifice! If you want a brighter glow, you can use a 40-W or 60-W incandescent lamp instead of a 25-W one (if you can still find any by the time you read this book).

CHAPTER 3

What Goes Where?

How does electricity get from the generating plant to your town, your neighborhood, and your house? You might wonder about that, the next time you have a local or regional power failure! You should also know which fuses or breakers affect which circuits and appliances inside your home or business. That knowledge can help you avoid overloading circuits, and to figure out how many devices or appliances you can safely use in a single outlet or circuit. If you travel to another country, can you use your cell-phone charger, your notebook computer, or your electric hair dryer there? You'll have a happier trip if you find out *before* you destroy something, not afterwards!

Whence the Juice?

Electrical energy morphs multiple times from birth to demise. Nevertheless, the initial source always exists as kinetic energy in some nonelectrical form, such as falling or flowing water (hydroelectric), coal or oil or methane gas ("fossil fuels"), radioactive substances (nuclear), moving air (wind), light from the sun (solar), or heat from the earth's interior (geothermal).

The Sources

In fossil-fuel, nuclear, geothermal, and some solar electric generating systems, heat boils water to make steam that passes through turbines under high pressure. The turbines drive massive electric generators. As the power demand increases, it takes more and more force to turn a generator shaft. That's why the utility companies need so much oil, coal, or gas to run a fossil-fuel power plant. Nuclear energy systems pretty much get rid of the fuel supply conundrum, but they produce radioactive waste that brings a whole new set of problems. Wind, solar, and geothermal power plants produce no waste or pollution when they operate, except for a little bit of residual heat energy. Hydrogen fusion power plants, if and when engineers manage to deploy them, will produce no waste other than heat and water vapor.

In a hydroelectric power plant, waterfalls, tides, or river currents directly drive specialized turbines that turn the generator shafts. In a wind-driven system, moving air operates devices similar to windmills, producing torque that turns the generator shafts. Although these types of power plants do not pollute the environment directly, they nevertheless present problems. The construction of a large hydroelectric dam can disrupt ecosystems, adversely affect agricultural and economic interests downriver, and displace people upriver by flooding their land. Many people regard arrays of windmill-like structures as an eyesore, but in order to generate significant electrical power, many such devices must be connected together and operated simultaneously. If you've ever driven past a large "wind ranch," you know that wind turbines can dominate the landscape.

In a photovoltaic (PV) energy generating system, semiconductor devices convert sunlight directly into DC electricity at low voltage. This DC must undergo conversion to high-voltage AC for transmission and distribution. The PV cells can't collect any energy during the hours of darkness, so storage batteries are necessary if a stand-alone PV system is to provide useful energy at night. Photovoltaics without storage batteries can work in conjunction with existing utilities to supplement the total energy supply available to all consumers in a power grid. A "solar ranch" has a lower profile than a "wind ranch," a feature that some people appreciate, especially those who live in the vicinity where wind turbines can ruin their views of the countryside.

Did You Know?

Whatever fuel type is used to provide the electricity that comes from a large power plant, the output current, which travels along those wire-and-tower "highways" that we can all instantly recognize, gets its impetus from electrical *potentials* (voltages) on the order of several hundred thousand volts, and in some cases, more than a million volts.

The Journey

When electric current travels along wires over great distances, some power goes to waste because of the wire resistance. This phenomenon cannot be avoided. No wire forms a perfect conductor, so we always end up losing some power in transmission lines because of "electrical drag." Engineers do their best to minimize this power loss. Two measures are commonly employed to that end.

First, engineers try to keep the wire resistance to a minimum by using large-diameter wires made from metal having excellent conductivity, and by routing the power lines in such a way as to keep their spans as short as possible. This approach can be carried out only to a certain extent in practice before physical and financial limitations "put a lid on it." No one wants to see a high-voltage utility line run right through a residential neighborhood in their town. Mountains, canyons, and lakes present a direct physical challenge.

Second, engineers use the highest possible voltage in a long-distance transmission line. Given a constant power demand, the current goes down as the voltage goes up.

The wasted power in the line varies according to the square of the current, so maximizing the voltage will minimize the power loss. These high voltages are common in cross-country *high-tension* (high-voltage) power lines.

The Destination

High-voltage electricity, despite its advantages for long-distance transmission, would never work in your house. It would instantly destroy your appliances, it would create a fire risk, and electrical arcs, like miniature lightning bolts, would kill you before you could put a plug into an outlet.

Did You Know?

The gigantic insulators that you see in high-voltage utility lines exist for a good reason. Under some conditions, a "spark" can jump several feet through the air at "high-tension" voltages. At best, an event like that wastes a lot of power. At worst, it can start fires and electrocute people or animals. The insulators ensure that the electricity stays in the wires and does not "jump out" where it doesn't belong.

Step-down transformers reduce the voltage of high-tension transmission lines down to a few thousand volts for distribution within municipalities. These transformers are physically large (about the size of a big car or small truck) because they must carry significant power. Several of them might be placed in a building or a fenced-off area. The outputs of these transformers feed local power lines that run along city streets.

Smaller transformers, usually mounted on utility poles or underground, step the municipal voltage down to 234 V for distribution to individual homes and businesses. These transformers are about the size of a trash barrel. Some utility outlets are supplied directly with 234 V. Large appliances, such as electric stoves, ovens, and laundry machines, commonly work at this voltage. In the United States and many other countries, smaller wall outlets and light fixtures receive single-phase electricity at 117 V.

Quick Question, Quick Answer

- Do you wonder why the electric utility companies produce AC instead of DC? Wouldn't DC work just as well, or better? Isn't DC simpler, after all?
- In theory, AC may seem more complicated than DC. But in practice, AC is easier to implement when you want to provide electricity to a lot of people spread over a large geographic area. To serve millions of consumers, you need the power of falling or flowing water, the ocean tides, wind, fossil fuels, controlled nuclear reactions, or geothermal heat. All of these energy

sources can drive turbines that turn AC generators. Second, AC lends itself to voltage transformation, a necessity for efficient transmission and distribution. It's more difficult to transform DC voltages.

For Nerds Only

Thomas Edison favored DC over AC for power transmission before the electric infrastructure had been designed and developed. His colleagues argued that AC would work better. But Edison knew something that his contemporaries apparently preferred to ignore, if they knew it at all. At extremely high voltages, DC travels more efficiently over long distances than AC does. Long lengths of wire exhibit less *effective resistance* (also called *ohmic loss*) with DC than with AC, and less energy goes to waste in the form of magnetic fields surrounding the wire. Direct-current high-tension transmission lines might, therefore, prevail someday, if engineers can find a way to bring the cost of such systems to within reason. Then Edison will have won a belated victory!

Phasing Schemes

Once the utility voltage has been stepped down to a level that won't zap everything in your home or business, the electricity is ready to be tailored for end use. Part of that process involves separating out the different *phases* of current. Phase is a fancy term that refers to points along an AC wave cycle. Phase can also express the extent of the timing difference between two AC waves that have the same frequency.

Degrees of Phase

You can specify time points in an AC wave by dividing one complete cycle into 360 equal parts called *degrees* or *degrees of phase*. Assign 0 degrees (0°) to the point in the cycle where the wave crosses the time axis going upward (with the voltage getting more positive). Then:

- One-quarter of the way through the cycle equals 90°
- Halfway through equals 180°
- Three-quarters of the way through equals 270°
- The end of the cycle equals 360°

Figure 3-1 illustrates this concept for a sine wave of the sort that you'll find at a household utility outlet. If you have trouble with the idea of dividing a wave into 360 degrees, imagine each cycle of the wave as a complete trip around a perfect circle. Then you can simply take advantage of a paradigm with which we're all familiar:

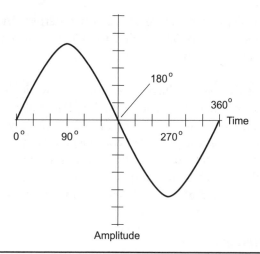

Figure 3-1 Degrees of phase tell you how much of a wave cycle has passed since its starting time. You can think of one complete wave cycle as going exactly once around a full circle.

- One-quarter of the way around a circle equals 90°
- Halfway around equals 180°
- Three-quarters of the way around equals 270°
- Exactly once around equals 360°

Waves in Phase Coincidence

The term *phase coincidence* means that two waves with the same frequency begin at the same instant in time. If the waveforms have identical shapes (but maybe different voltages), they follow each other along from instant to instant. Figure 3-2 shows

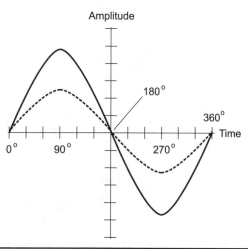

Figure 3-2 Here's a graph of sine waves in phase coincidence; the phase difference in this case equals 0°. The two waves follow right along with each other.

an example of phase coincidence between two perfect sine waves whose voltages differ. The *phase difference*, also called the *phase angle*, equals 0°. In this kind of situation, you always know several things:

- The positive peak voltage of the resultant wave, which is also a sine wave, equals the sum of the positive peak voltages of the two composite waves.
- The negative peak voltage of the resultant wave equals the sum of the negative peak voltages of the composite waves.
- The peak-to-peak voltage of the resultant wave equals the sum of the peak-to-peak voltages of the composite waves.
- The phase of the resultant wave coincides with the phases of the two composite waves.

Waves of Differing Phase

Two perfect sine waves having the same frequency can differ in phase by any amount from 0° (phase coincidence), through 90° (*phase quadrature*, meaning a difference of 1/4 of a cycle), through 180° (a difference of half a cycle), through 270° (phase quadrature again, but a difference of 3/4 of a cycle), and finally 360° (phase coincidence, but offset by a full cycle).

Waves 180° out of Phase

When two pure sine waves of identical frequency begin exactly half a cycle apart in time, engineers say that they occur *180° out of phase* with respect to each other. Figure 3-3 illustrates a situation of this sort.

If two perfect sine waves have identical voltages and occur 180° out of phase, then they cancel each other out. The voltages of the two waves are equal and

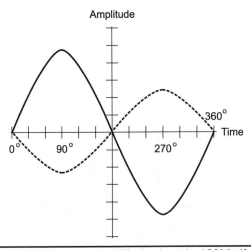

FIGURE 3-3   Here's a graph of sine waves that differ in phase by 180° (half a cycle).

opposite at every point in time, so they always add up to zero! If two perfect sine waves have different voltages and occur 180° out of phase, then:

- The peak-to-peak voltage of the resultant wave, which is also a sine wave, equals the difference between the peak-to-peak voltages of the two composites.
- The phase of the resultant wave coincides with the phase of the stronger of the two composite waves.

Did You Know?

You can define the *phase relationship, relative phase, phase difference,* or *phase angle* (all four terms mean the same thing) between two waves only when the two waves have the same frequency. If the two waves have different frequencies, then their relative phases change constantly, and you can never "nail it down" to any particular value.

For Nerds Only

Moving a wave forward or backward by half a cycle (changing its phase by 180°) doesn't necessarily alter the wave in the same way as *phase opposition* (turning the wave upside down). A pure sine wave gives you a special case where the two actions produce the same practical result in ordinary electrical circuits. However, with other waves, the results usually differ. To see these effects, draw graphs of some irregular or lopsided waves, and then compare the results of a 180° phase shift with phase opposition.

Phase Options

Single-phase AC consists of a single, pure sine wave. You'll find this sort of AC at standard wall outlets intended for small appliances, such as lamps, TV sets, and computers. In most parts of the United States, the voltage is standardized at 117 V RMS for single-phase AC, but it can vary a few percentage points above or below this level, depending on the overall utility power demand at the time, your location, and the whims of your local electric power provider.

Over long-distance power lines, utility companies transmit electricity in the form of three sine waves, each having the same RMS voltage, but differing in relative phase by 120° (1/3 of a cycle). Engineers call it *three-phase AC*. Each sine wave travels along its own wire, so the transmission line has three wires (or pairs of wires to minimize resistance losses). In addition, a well-designed transmission line always has a single wire connected to a good electrical ground, placed above the current-carrying lines. This grounded wire serves to "attract lightning away" from the power lines. Lightning behaves in unpredictable ways, but it tends to strike objects connected to a good electrical ground, and it also tends to strike the highest or tallest thing in the vicinity.

Figure 3-4 shows three-phase AC as a graph of voltage versus time. The three individual waves are called *phase 1*, *phase 2*, and *phase 3*. The horizontal axis increases in degrees of phase as you go from left to right, based on the start of the wave for phase 1 (shown as a solid curve) where the voltage equals zero and increases positively. Phase 2 (shown as a dashed curve) comes 1/3 of a cycle later than Phase 1, and phase 3 (shown as a dotted curve) comes 2/3 of a cycle later than phase 1. Therefore, phase 2 starts its cycle 120° later than phase 1 starts, and phase 3 starts its cycle 120° later than phase 2 starts.

Did You Know?

Three-phase AC travels more efficiently along power lines than single-phase AC does. As a result, less energy goes to waste as heat in the wires. The electromagnetic fields from three phases, each one traveling along its own wire in a closely spaced arrangement, mutually cancel each other out at points far away from the transmission line. That effect prevents "radiation loss" from the line. If long-distance power transmission were done as single-phase AC, the span of wire would act like an *extremely-low-frequency* (ELF) antenna, radiating valuable energy out into the surrounding environment where it could do no good for end-of-the-line users.

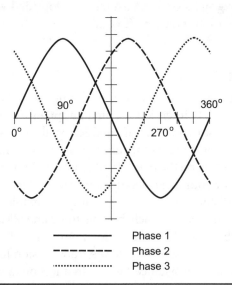

Figure 3-4   Here's a graph of three-phase AC. Each pure sine wave is separated by 120°, or 1/3 of a cycle, from the other two.

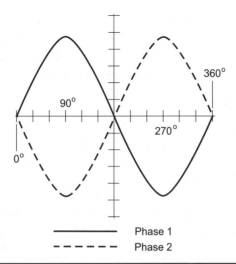

FIGURE 3-5   Here's a graph of split-phase AC. The two pure sine waves are separated by 180°, or half a cycle, so they directly oppose if used together, in effect doubling the RMS voltage of either wave alone.

For residential and business service in most of the United States and Canada, you'll find *split-phase AC* in common use. Two sine waves, each at 117 V RMS, travel along their own dedicated wires, with a third wire connected to electrical ground. Figure 3-5 shows a graph of the waves in this arrangement. The two phases directly oppose each other. Households can take advantage of either phase alone, along with the grounded wire, to serve 117 V RMS appliances, such as lamps, computers, and television sets. Both phases can be used together so that they *buck* (work against) each other, resulting in an effective voltage of 234 V RMS between them (234 V equals twice 117 V).

When you look closely at the utility wires coming into your house (assuming that they're above ground so you can see them), you'll probably find two black insulated wires wrapped around a single bare wire. The bare wire goes to electrical ground. The two insulated wires carry the 117 V RMS sine waves in split-phase form. Your fuse or breaker box separates the phases. Some parts of your house receive 117 V RMS from one phase; other parts of the house get 117 V RMS from the other phase. Heavy-duty appliances, such as laundry machines and electric ovens, receive both phases bucking each other along with an electrical ground, providing 234 V RMS, the effective voltage difference between the two opposing phases.

Your Electric Meter

An electric meter measures the amount of energy that your household consumes in *kilowatt hours* over a period of time. A kilowatt hour (1 kWh) equals the amount of energy that a 1000-W (one-kilowatt or 1-kW) appliance consumes in an hour, or the equivalent of it. For example, a 100-W incandescent bulb will take 10 hours to consume 1 kWh, a 10-W compact fluorescent lamp (CFL) will take 100 hours to consume 1 kWh, and a 1-W light-emitting-diode (LED) lamp will take 1000 hours to use up 1 kWh. You can express energy as power that accumulates as time passes. Conversely, you can express power as energy consumed per unit time.

Mechanical Meters

An analog mechanical electric meter employs a small motor, the speed of which depends on the current, and therefore, on the power at a constant voltage. The number of turns of the motor shaft, in a given length of time, varies in direct proportion to the number of kilowatt hours consumed. The motor connects at the point where the utility wires enter the building. That's where the utility system splits into three circuits: one providing 234 V for heavy-duty appliances, such as the range, oven, washer, and dryer, and two others providing 117 V for small appliances, such as lamps, computers, and television sets.

If you observe one of these old-fashioned kilowatt-hour meters, you'll see a disk spinning: sometimes fast, sometimes slowly. Its speed at any particular moment in time depends on the power that your household is using at that moment. The total number of times that the disk goes around, hour after hour, day after day during the course of each month, determines the size of the bill that you get from the power company for that month. Your bill also depends on how much the electric company charges you for each kilowatt hour that you use, of course.

An analog electric meter has several scales calibrated from 0 to 9 in circles, some going clockwise and others going counterclockwise. To read the device, you must make your "mind's eye" go in whatever direction (clockwise or counterclockwise) the scale goes for each individual meter. Figure 3-6 shows an example. You read the dials going from left to right. In this example, the leftmost dial rotates counterclockwise, the second one goes clockwise, the third one goes counterclockwise, and the fourth one (the one on the far right) goes clockwise. For each meter scale, you can write down the number that the pointer has most recently passed. In this case those values are 3, 8, 7, and 5, so you would write 3875; and by looking closely at the far-right-hand dial, you can surmise that the meter indicates slightly more than 3875 kWh.

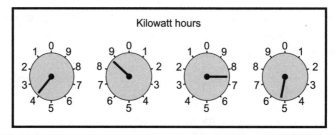

Figure 3-6 A mechanical electric meter with four rotary dials. This meter displays a total consumed energy amounting to slightly over 3875 kWh.

Smart Meters

In some locales, electromechanical meters of the sort shown in Fig. 3-6 have been supplanted by devices that provide a direct digital readout. The simplest ones replace the rotary dial pointers with rotating cylindrical drums that have numerals printed on them. That allows you (and the meter reader) to simply write down the digits.

More sophisticated electric meters, which have proliferated in recent years, contain no moving parts. All of the mechanical components have been replaced by solid-state electronic devices and displays. Some such meters can do a lot more than simply keep track of your energy consumption. They can interconnect with supplemental systems, such as solar panels and wind turbines, allowing you to reduce your monthly electric bill, and in some cases, make a profit by selling surplus power to the electric company. Meters of this type are called *smart meters*.

A smart meter can register how much power (in kilowatts) you use, on the average, for specific intervals in time (usually 15 minutes) and keep a record of it. You can see the information on recent usage by watching your smart meter's display for a minute or two. One of the readings will show a numeral followed by "kW," which quantifies power (instead of "kWh" which quantifies energy). If you run an air conditioner, an electric oven, an electric water heater, and a refrigerator all at the same time, you'll see a large number there. If you leave your house for a few days, shut everything down, and then look at the meter immediately when you return (and before you switch anything on again), you'll see a small number there.

Fact or Myth?

Have you heard that smart meters can tell the power company (or the government) what appliances you use, and when you use them? Well, that's partly true, but not completely, and in any case, it doesn't tell you the whole story. While peak-power numbers can let the electric company know how much you consume at a maximum, those values don't divulge what specific appliances you use, or when you use them. Theoretically, the electric company could use your smart meter to shut your house down, but they're not likely to do that unless a dangerous event, such as a gas leak, occurs where an electrical spark could start a fire or cause an explosion. In that case, you'll be happy that the electric company is willing and able to take measures that could save your property or even your life!

The Perfect Breaker or Fuse Box

Have you ever seen a breaker box or fuse box where all the circuits have correct labels, telling you exactly which parts of the house each breaker or fuse protects? I haven't ever come across such a thing. Even my own breaker box, which I painstakingly labeled a few years ago, still has mistakes. It's as if gremlins lurk in the ether, waiting to sabotage our hapless attempts to create The Perfect Breaker or Fuse Box (TPBOFB).

Fixed Appliances

Only one method exists that can guarantee the attainment of TPBOFB: Individually test every installed light fixture, every installed appliance, and every electrical outlet in the house against the breakers or fuses in the box. Permanent light fixtures and installed appliances, such as electric ovens, electric laundry machines, central air conditioners, and furnace fans, lend themselves easily to such tests. You need only switch them on and then open up your breakers or remove your fuses, one by one, until the relevant device goes off.

Warning!   Always wear your gloves and shoes when you fiddle around with a breaker box or fuse box. You never know when the "Wicked Wizard of Watt" will try to clobber you. But he can't hurt a well-protected person!

In the case of a permanent light fixture of the sort that you find on ceilings or exterior walls, you must make sure that the switch controlling the fixture stays on while you conduct the tests. (That rule should seem obvious, but it's easy to forget.) Switch the light on and leave it on while you work with the breakers or fuses. Also, make sure that every permanent light fixture has a functioning bulb in it. The same rule applies for any and all wall outlets controlled by switches. Turn all such switches on, and leave them on for the duration of the test.

An Outlet Tester

You can construct a simple electrical outlet tester with a light-emitting-diode (LED) lamp and a socket with a standard lamp base and two prongs that go into 117-V outlets. The best LED for this purpose is one of the sort shown in Fig. 3-7. It's

FIGURE 3-7   You can assemble an outlet tester with a standard plug-in bulb socket and a light-emitting-diode (LED) lamp.

physically rugged, it will last for a long time, it'll shine brightly enough to see in daylight but not so bright as to create a distraction, and it's not so big that it'll burn you when you pull it out of the socket. Screw the lamp into the plug-in socket and test it on a live outlet to make sure that it works.

You can use your outlet tester to check any outlet within reach as you ramble between the breaker or fuse box and watch for "glow" vs. "not glow" conditions! If you're too lazy to walk back and forth through your house dozens of times, you can employ the services of a child, spouse, or other willing person, and communicate with each other using your cell phones. Then you can carry out a variant of the famous cell-phone classic conversation: "Is it lit now?" (pause) "Is it lit now?" (pause) "Is it lit now?" and so on.

Make a Blueprint!

If you want to create TPBOFB (or try), you'll have to find a blueprint of your house that shows each and every outlet and electric light fixture. Then label each outlet and fixture with the appropriate number in your breaker or fuse box. Keep a copy of that blueprint right next to the box.

Alternatively, you can make up a detailed list with two columns: full descriptions of each and every outlet and fixture in the left-hand column, and the corresponding breaker or fuse number in the right-hand column. Keep the list stored in your computer, so that you can edit it as new, never-before-known outlets appear in your house, or as outlets decide to change from one breaker or fuse circuit to another. (I'm only kidding about this stuff—I hope!)

Did You Know?

As for the merits of having TPBOFB in your house, rest assured that it'll come in handy some day. Maybe you won't need it this week, maybe not this month, maybe not this year, but eventually you'll be glad that you went to the trouble to create it.

Circuit Management

In American residential homes, the individual 117-V circuits are usually rated for a maximum current of 15 or 20 amperes (or "amps," abbreviated A). Once in awhile you'll see a circuit rated at 30 A. The 234-V circuit (or circuits), if any exist, are rated at 20 to 50 A. When you use a lot of heavy appliances, you must keep these limitations in mind.

Loading Them Up

Table 3-1 lists several common types of household electrical appliances along with the amounts of current that they draw in a 117-V circuit. As a general rule, you

TABLE 3-1   Amounts of Current Drawn by Various Common
Appliances in 117-V Residential Circuits

Appliance	Current in Amperes
60-W-equivalent LCD lamp	0.043
60-W-equivalent CFL	0.12
15-W incandescent lamp	0.13
40-W incandescent lamp	0.34
100-W incandescent lamp	0.85
Slow-cook crockpot set on "low"	0.6–1.0
Slow-cook crockpot set on "high"	1.3–2.0
Electric roaster kettle	7–10
Electric roaster oven	13–14
Small electric fan	0.5–1.5
Large electric fan	1–2
Electric space heater set on "low"	5–11
Electric space heater set on "high"	13–14
Small window air conditioner	8–12
Electric frying pan	10–13
Laptop computer	0.5–1.0
Desktop computer and LCD display	3–5
Big-screen TV set with LCD display	3–5
Inkjet printer	1–2
Laser printer	5–10

should try to avoid going over the 2/3 point with any individual circuit in your house (unless you want to use a large appliance, such as an electric space heater or electric frying pan, of course). That means, for example, that you should avoid trying to load down a 15-A circuit with more than about 10 A, or a 20-A circuit with more than about 13 A. Once in awhile you'll have to exceed this limitation, but you should stay within it as much as you can.

If your main electric distribution box has circuit breakers, the breakers will "automatically" force you to abide by their limitations. You'll usually get a little bit of time, if you "max out" a particular circuit, before a breaker trips, unless you overload the circuit severely or subject it to a "dead short." (Once, I made the mistake of using a diagonal cutter to sever a live electrical cord. The breaker tripped instantly. Fortunately, the cutter had insulated handles, and I was wearing gloves!) In the case of a fuse box, you'll have less time. If and when a fuse blows out, you should always replace it with another fuse of the same type and the same current rating.

Warning! Never replace a blown fuse with anything other than the same type, having the same current rating. If you try to "cheat" by replacing, say, a 15-A fuse with a 20-A fuse, you'll risk overheating the wires in your home, a notorious cause of electrical fires. If you replace a fuse with one having a smaller current rating, say a 15-A fuse instead of a 20-A fuse, you'll be perfectly safe, although you might find the stricter limitation inconvenient.

Extension Cords and Block Taps

Extension cords come in a great variety of lengths, with different wire sizes (or gauges) depending on how much current they can safely carry before they get too hot and pose a fire hazard. Figure 3-8 shows two common extension cords. The one on top measures 40 feet (approximately 13 meters) long and has three wires, while the one on the bottom is 6 feet (about 2 meters) long and has two wires. The shorter cord has a three-outlet block on the end, allowing for the connection of multiple appliances. Both cords are rated for a maximum load of 15 A at 117 V.

Figure 3-9 shows a *block tap* that you can plug into a dual-outlet wall receptacle in order to use more than two appliances with that outlet. In this example, only two appliances are connected to the entire circuit, so technically the block tap isn't needed. You should avoid using block taps in conjunction with high-current appliances, such as electric heaters, frying pans, or roaster ovens.

Figure 3-10 shows a block tap used incorrectly with an extension cord. The three-outlet tap goes into one of the outlets in the extension cord that you saw in Fig. 3-8. This "tap-on-a-tap" scheme has an extra connection where the prongs of a plug contact the slots of an outlet. Every time you create a temporary connection like that, you run the risk of overheating, particularly with appliances that draw a lot of current.

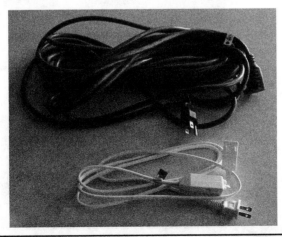

Figure 3-8 A 40-foot, three-wire extension cord for indoor/outdoor use, and a 6-foot, two-wire extension cord with three outlets for indoor use. Both cords are rated at 15 A.

FIGURE 3-9   Proper use of a block tap in a 117-V electrical outlet.

Warning!   Never connect multiple load-splitting devices right next to each other in cascade (one after the other). If you fill up all of the available outlets in a contraption such as the one shown in Fig. 3-10, you'll run the risk of overheating the splitters because they're too close together. That will create a potential fire hazard, even if you don't strain the circuit's breaker or fuse. Also, if you use a high-current appliance in this arrangement (even a single one), you'll create a fire hazard.

FIGURE 3-10   Improper use of a block tap. Never cascade load-splitting devices one after another like this!

"Tanglewire Gardens"

Most of us have computer workstations, and some of us have lots of peripherals and ancillary equipment, such as a printer, a scanner, a modem, a router, a cordless phone, a desk lamp or two, a charging bay (for devices such as tablet computers and cell phones), and so on. All of these things get their power, either directly or indirectly, from the 117-V utility system. As a result, anyone with a substantial computer workstation will end up with a mess of wires behind and underneath the work desk: a "tanglewire garden"!

"Tanglewire gardens" can look dangerous, as if they would inevitably present a high fire risk, but they don't have to pose a hazard. If you know how to connect and arrange the wires properly, it doesn't matter from a safety standpoint how much you tangle them up, although you might want to affix labels on the cords near their end connectors (on each end) so that you don't get them confused with each other when the inevitable malfunction occurs and you have to pull out and replace one of the components of your system.

Figure 3-11 shows the "tanglewire garden" underneath my home electronics workbench. In addition to a computer, this system includes an amateur ("ham") radio transceiver, two displays, an interface between the radio and the computer, a microcomputer-controlled power-measuring meter, an audio amplifier for the computer and radio, a wireless headset, a desk lamp, and an external hard drive that needs its own "power brick." That's 10 devices or cords in total, all deriving their power from a single outlet in the wall underneath the table!

Figure 3-11   "Tanglewire garden" beneath the author's electronics workbench. A heavy-duty UPS (out of the picture to the right) serves two power strips mounted on a metal baking sheet that rests on detached plastic shelves.

In order to ensure smooth operation of the system in case of a power failure, all of the devices are connected to the wall outlet through a commercially manufactured *uninterruptible power supply* (UPS). The UPS has a battery that charges from the AC utility under normal conditions, but provides a few minutes of emergency AC (with the help of some sophisticated electronic circuits) if the utility power fails. That few minutes gives me time to deploy my backup generator without having to shut any of the devices down. The UPS has four outlets in the back, two of which go to power strips with six outlets each, and the other two of which remain empty. There are 12 available outlets in the power strips, 10 of which are engaged, as shown in Fig. 3-11. The UPS also has a transient voltage suppressor built-in. Figure 3-12 is a block diagram of the general arrangement.

Did You Know?

You should not use power strips with transient suppressors in conjunction with any other component, such as a UPS, that also has a transient suppressor. When you connect two or more transient suppressors in series (cascade them one after another), they'll probably conflict. Therefore, in the system shown in Figs. 3-11 and 3-12, the power strips do not include transient suppressors (or, as ill-informed people call them, "surge protectors").

Quick Question, Quick Answer

- Doesn't the presence of 10 devices, all plugged into a single wall outlet, as shown in Figs. 3-11 and 3-12, create a danger by overloading the outlet?
- Not in this case! All of the devices, taken together, consume less than 10 A (2/3 of the breaker rating for the outlet), even if they all run at once. The radio interface, the cordless headset, the audio amplifier, the power-measuring meter, and the desk lamp draw less than 1 A combined. The rest of the devices, taken together, draw about 7 A.

I've taken three extra precautions, aside from making sure that I don't overload the wall outlet, to ensure that my "tanglewire garden" remains safe. You should do the same with your pride and joy!

1. First, if you look carefully at Fig. 3-11, you'll notice that I've mounted the power strips on a metal sheet. It's a solid aluminum baking sheet. I glued the strips down there with epoxy resin. This precaution keeps the power

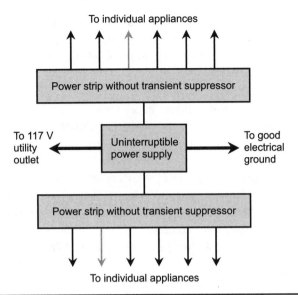

To individual appliances

To 117 V utility outlet

Power strip without transient suppressor

Uninterruptible power supply

To good electrical ground

Power strip without transient suppressor

To individual appliances

FIGURE 3-12 Block diagram of the "tanglewire garden" beneath the author's workbench. The power strips include breakers but not transient suppressors; the UPS contains a transient suppressor that serves the whole system. Gray arrows represent unused outlets.

strips from setting anything (other than themselves) on fire if they start sparking, a stunt that these things have been known to perform, occasionally with disastrous results.

2. Second, I have not allowed any cord splices or other sensitive electrical points to lie directly on the floor. The baking sheets, as well as all points in the cords where splices exist, are set up on thick plastic shelves. My basement floor will get wet if a huge, sudden rainstorm occurs. (Of course, in that event I won't use the workstation until the floor dries out!)

3. Third, I've connected a dedicated ground wire from the chassis of the UPS to a known electrical ground. I tested the wall outlet underneath the workbench to ensure that the "third prong" actually goes to the electrical ground for the entire house. You can test the "third prong" of any three-wire outlet by following the procedure I describe in "Grounded, or Not?" a little later in this chapter.

Which Circuit, Which Phase?

You can use a long extension cord, along with your multimeter, to find out whether two 117-V circuits are on the same phase or not. If they're on the same phase, the

AC voltage between the two live outlet slots (shorter ones) will equal 0. If they're on opposite phases, you'll see 234 V, plus or minus a few percentage points, between them. Perform these steps in order, using Fig. 3-13 as a reference, and watch the results.

1. Put on your rubber gloves and shoes.
2. Consider this test only for three-wire outlets, never for two-wire outlets.
3. Decide which pair of outlets you want to compare.
4. Plug a long three-wire extension cord into one of the outlets.
5. Position the outlet end of the extension cord next to the other outlet.
6. Set your multimeter for AC volts, with an upper range limit of 250 V or more.
7. Place the black probe tip of the meter into the shorter of the two rectangular slots in the extension cord outlet.
8. Place the red probe tip into the shorter of the two rectangular slots in the wall outlet.
9. Read the meter.
10. If the meter says 0 V, then the AC waves of the two outlets coincide in phase (in other words, they follow along with each other, as shown in Fig. 3-14A).
11. If the meter says 234 V (or something near it), then the AC waves of the two outlets oppose in phase (in other words, they buck each other, as shown in Fig. 3-14B).

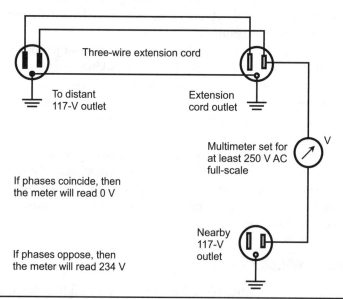

FIGURE 3-13   Arrangement for determining whether the phases of two 117-V utility circuits coincide or oppose. Make sure to set the multimeter to measure AC volts, with a full-scale rating of at least 250 V. Wear your gloves at all times.

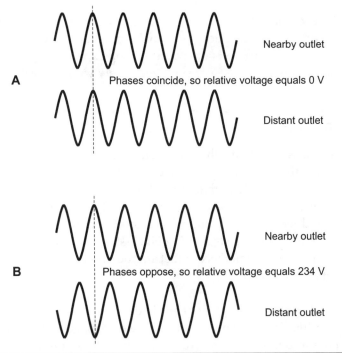

Nearby outlet

A   Phases coincide, so relative voltage equals 0 V

Distant outlet

Nearby outlet

B   Phases oppose, so relative voltage equals 234 V

Distant outlet

FIGURE 3-14   At A, the electrical waves follow along with each other, so the two outlets have no
voltage difference relative to each other. At B, the electrical waves oppose each
other, so the two outlets have a difference of 234 V AC.

Quick Question, Quick Answer, and a Warning!

- You might wonder whether or not you can combine two or more different in-phase circuits to increase the available current from a single 117-V circuit. Can you connect them in tandem (parallel)? Can you combine two out-of-phase 117-V circuits to operate a 234-V appliance, such as a washing machine or an electric range? In other words, can you make a brand new 234-V circuit by tapping the short slots of two different outlets that work in phase opposition?

- In theory, you can do either of these things. But in practice, you'd better not! If you want to obtain a circuit with a higher current rating than any of the ones in your house can provide, or if you want to obtain a 234-V circuit when your house doesn't already have one, have a professional electrician do the wiring and install everything according to established electrical codes. Don't try to do it yourself, and above all, never fool around with multiple circuits in your home wiring system. If you try to "brew your own super-circuit" this way, not only will you run the risk of fire or electrocution, but you'll probably void your homeowner's insurance policy to boot. The foregoing experiment is intended only for educational purposes, or simply to satisfy your curiosity.

Grounded, or Not?

Many people assume that a three-wire 117-V utility outlet has a good ground at its "third hole" (the bottom hole, not either of the vertical slots). That's not always true! I've seen cases where that "third hole" wasn't connected to anything at all. You can use a long extension cord and your multimeter to find out whether or not a particular three-wire 117-V outlet has a good electrical ground at its "third hole." Go through the following steps using Fig. 3-15 as a reference. For this test to work, you'll need to find a reference outlet in your house that you know has a good ground at its "third hole."

1. Put on your rubber gloves and shoes.
2. Consider this test only for three-wire outlets, never for two-wire outlets.
3. Plug a long three-wire extension cord into an outlet with a known good electrical ground at the "third hole."

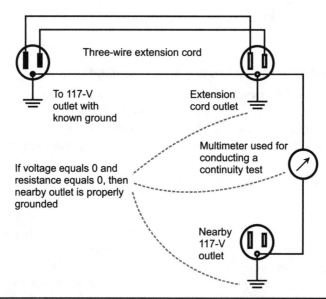

FIGURE 3-15 Arrangement for determining whether the "third prong" of an outlet is properly grounded. Follow the procedure for a continuity test. Always check for voltage before measuring resistance! Wear your gloves at all times.

4. Position the outlet end of the extension cord next to the outlet whose "third hole" you want to test.

5. Set your multimeter to measure the highest AC voltage that it can deal with.

6. Insert the multimeter's black probe tip into the "third hole" in the extension cord outlet.

7. Insert the multimeter's red probe tip into the "third hole" of the outlet you want to test, while leaving the black probe tip in the extension cord outlet.

8. Check the meter reading. If it shows anything other than 0, then an AC voltage exists between the two points, so your outlet *does not* have a good ground. It will present a shock or fire hazard if you use it.

9. If you see 0 as the result for step 8, switch the meter to the next lower AC voltage function and repeat the test.

10. If you get 0 again, repeat steps 8 and 9 with all the AC voltage functions that the meter has, going down until you've tested at the lowest AC voltage setting. You should always get a reading of 0. If you don't, then you know that some AC voltage exists between the two points, so your outlet *does not* have a good ground.

11. Assuming that you've seen readings of 0 for all the AC voltage settings, repeat the tests with your multimeter's DC voltage settings, starting with the highest one and working your way down to the lowest one. Test for DC voltage in both directions: first with the black probe tip in the "third hole" of the cord outlet and the red probe tip in the "third hole" of the outlet under test, and then the other way around. You should always see a meter reading of 0.

12. If you ever see any DC voltage besides 0 between the two points, then you know that your outlet *does not* have a good ground.

13. If you see 0 for all of the AC and DC voltage results, switch your multimeter to the highest resistance function.

14. Touch the meter's two test probe tips to each other. If you're using an analog meter, tweak the "0Ω ADJ." knob so that the meter indicates 0. If you're using a digital meter, make sure that the display indicates a value of 0.

15. Insert the black probe tip back into the extension cord's "third hole," and the red probe tip back into the outlet's "third hole." You should get a reading of 0.

16. Reverse the test leads. You should again get a reading of 0.

17. Repeat steps 14 through 16 for all the rest of the meter's resistance settings, working your way down one setting at a time, until you get to the lowest one.

18. If you have observed readings of 0 for every step in the foregoing process, then you can have confidence that your outlet's "third hole" is properly grounded.

19. If you ever see any meter reading other than 0 for the condition between the two "third holes," you know that your outlet *does not* have a good ground. In that case, treat it as a two-wire outlet until you can get a professional electrician to wire it up properly.

Did You Know?

If the "third hole" in an outlet connects to a wire that isn't grounded, that wire can pick up stray AC by *electromagnetic induction* from other wires in your house, giving you a nasty shock. It can also cause sensitive electronic equipment to malfunction. In addition to all that, if you connect a transient suppressor into the circuit, it won't protect anything because it will have no ground to "work against."

For World Travelers

If you live in the United States, you've grown accustomed to well-regulated AC that ranges from about 110 to 130 V RMS at 60 Hz. (The nominal figure of 117 V is commonly used for all such utility power.) Heavy-duty appliances, such as washing machines, electric ranges, water heaters, and central air conditioners, often use twice that voltage, ranging from about 220 to 260 V (and which we, in this book, refer to as 234 V), also at 60 Hz. However, in many other countries, the electrical specifications differ from those in the United States. The most common voltage is 234 V (give or take a few percent), often at 50 Hz rather than 60 Hz.

When it comes to AC, the frequency rarely matters in everyday use. In most situations, 50 Hz will do the same job as 60 Hz will. The exception is an old-fashioned electric clock that keeps its time based on a 60-Hz line frequency. (It will run slow at 50 Hz.) However, you can't expect to plug a device designed for 117 V into a plug that delivers 234 V and get anything other than a catastrophe. Fortunately, the outlets in other countries usually differ geometrically from those in the United States, so you can't simply plug your electric razor into a wall outlet in, say, Russia. You'll have to buy an *outlet adapter*. Most good hardware stores stock outlet adapters. Radio Shack stores carry a good supply as well.

Before you travel to any other country (whether you live in the United States or not), check with a travel agent, and also with the authorities in that country, to make sure that you know what voltage to expect at the utility outlets. Also, make sure that you know what sort of outlet configuration you'll find. The best travel agents will be able to tell you exactly which type of adapter and voltage converter you'll need.

You can use the Table of Electricity at Utility Outlets in Various Countries throughout the World in the Appendix as a starting point to figure out what you should expect, but don't take this table as a final authority. As mentioned above, double-check with government authorities in the country of interest, and/or with a qualified travel agent, before purchasing your adapters. At the time of this writing, a comprehensive information guide for voltages and adapter plugs throughout the world was available at the website http://www.voltageconverters .com/voltageguide.htm.

Did You Know?

Some laptop computer "bricks" (chargers) will let you select between 117-V and 234-V sources. The switch might be labeled "110" and "220" or maybe "120" and "240." However, not all laptops have this feature.

Warning! In some countries or remote locations, the voltage can fluctuate considerably from day to day, hour to hour, and even minute to minute. A simple voltage transformer or outlet adapter can't do anything about such fluctuations.

CHAPTER 4

Alternative Electricity

Natural disasters can disrupt the power grid for days, weeks, or months at a time. Most of us aren't prepared for a long-term power blackout. We might think that the probability of such an event is so low that full preparedness doesn't justify the cost (until the worst-case scenario occurs). But what about a three-hour blackout in winter that leaves your house at 50 degrees Fahrenheit because your furnace fan couldn't circulate the air without electricity? Suppose that you want to get off the utility grid altogether, or generate enough of your own electricity so that the power company pays you for the surplus?

Small Combustion Generators

You can find compact, portable *combustion generators* for use in homes and small businesses. Some combustion generators are also suitable for use by campers. For people living in remote areas, a combustion generator might constitute the primary, if not the only, source of AC electricity for common appliances.

How They Work

A small combustion generator provides 117 V AC in the United States (234 V in many other countries). Larger generators in the United States also supply 234 V AC for heavy appliances, such as electric ranges and laundry machines. The generator's internal combustion engine can range in size from a few horsepower (comparable to the one in a lawn mower or snow blower) to hundreds of horsepower (comparable to the engines in trucks, tractors, and construction equipment). Most small generator engines burn gasoline. Larger ones burn diesel fuel, propane, or methane.

In a mechanical AC generator, a coil of wire, attached to the shaft of the combustion engine, rotates inside a pair of powerful magnets. If you connect a load (such as an appliance) to this coil, an AC voltage appears across that load as each point in the wire coil moves past the *lines of flux* produced by the magnets, first in one direction and then in the other direction, over and over. In an alternative arrangement, the magnetic poles revolve around the wire coil, which remains fixed.

The AC voltage that a generator can produce depends on the strength of the magnets, the number of turns in the wire coil, and the speed of rotation. The AC frequency in a simple generator depends only on the speed of rotation. In the United States, the speed is 3600 revolutions per minute (3600 r/min) or 60 revolutions per second (60 r/s), resulting in an output frequency of 60 cycles per second (60 Hz). In many other countries, the rotational speed is 3000 r/min, producing an AC frequency of 50 Hz. In order to maintain a constant rotational speed for the generator under conditions of variable engine speed, mechanical regulating devices are required.

When you connect a load to the output of a simple generator, the engine has a harder time turning the generator shaft, as compared with the situation when no load exists. As the amount of electrical power demanded from a generator increases, so does the mechanical power required to drive it, and therefore, the amount of fuel consumed per unit of time. The electrical power that comes out of a generator is always less than the mechanical power required to drive it. The lost energy shows up as heat in the generator components. To maintain the proper AC frequency, a simple generator's engine must run at a constant speed under conditions of variable load. This state of affairs can prove difficult to attain, but there's a way around it!

For Nerds Only

The *efficiency* of a generator equals the ratio of the electrical power output to the mechanical driving power, both measured in the same units, such as watts (W) or kilowatts (kW), multiplied by 100 to get a percentage. No generator operates at 100-percent efficiency, but a good one can come fairly close to that ideal.

Small Gasoline-Powered Generators

Advanced small-scale generators circumvent the need for constant motor speed by converting the generated AC to regulated DC, and then using a power inverter to generate AC from that DC. If the motor speed changes, the DC voltage stays the same because the regulator circuit holds it constant, so the output AC voltage stays constant too. In the best commercially manufactured generators, the inverter produces a near-perfect sine wave to ensure that the machine can properly operate sensitive electronic devices, such as computers, printers, scanners, modems, and routers. A "raw" generator will produce a facsimile of a sine wave, but not of the quality needed by microcomputer- and microcontroller-based devices in common use today.

Figure 4-1 shows a popular portable gasoline generator with a power inverter that can provide up to 2 kW of clean sine-wave AC electricity when needed. This machine can run any of my computers, microcomputer-controlled furnace, and microcomputer-controlled amateur ("ham") radio transceivers perfectly well. It has a tank that holds 1.1 gallons (4 liters) of high-octane gasoline. With a load of a

FIGURE 4-1   A portable gasoline-fueled generator, capable of providing up to 2 kW of clean sine-wave AC power at 117 V RMS. The tied-up cord is the ground wire.

few hundred watts, that amount of gasoline provides several hours of continuous, reliable AC electricity. This generator has proven itself worthy as a backup power source in winter storms when utility failures would otherwise have meant no heat for my house, as the furnace electronics and fan require 117 V AC to function!

Did You Know?

Any backup generator, if poorly designed, can cause problems if you try to run sensitive electronic equipment from it. However, a well-engineered generator with a power inverter, even the small gasoline-fueled type, will work fine with computers and other sophisticated systems as long as you keep it in proper working order. If you want your generator to be available when you need it, you must adhere to a maintenance schedule that involves cleaning, spark-plug replacement, and periodic testing.

Quick Question, Quick Answer

- How large can home-based and business-based generators get, from an availability standpoint?
- Medium-sized diesel-, propane-, and methane-fueled generators can supply several tens of kilowatts, and can power an entire home, business, or agency. Large institutions typically have two or more generators. These machines, if properly operated and maintained, can operate all kinds of equipment, even sensitive and complex medical devices.

My Arrangement

A Honda EU-2000i portable gasoline-fueled generator (Fig. 4-1) forms the heart of my emergency backup power ensemble. In addition to the generator, I use several extension cords and power strips to distribute electricity to the points where I need it the most during a utility outage. I always keep in mind the maximum power that the generator can provide; I never let it come close to "maxing out" at the full 2-kW limit. You can use this general configuration as a template for your own system, if you want to install one, tailoring the specifics to meet your needs.

Did You Know?

Honda, Yamaha, and other manufacturers offer several portable generator models, some of which are a little smaller than mine, and some of which are quite a lot bigger. Whatever brand of generator you decide to buy, you should make certain that it produces a "clean" sine wave for sophisticated electrical and electronic devices. Always check not only with the generator dealer (who will tell you the truth in a perfect world, but in the real world, maybe not), but also with the manufacturer's specification sheet. E-mail or telephone the manufacturer and ask for details about the model that interests you. Remember that the best generators employ power inverters to produce clean, regulated, sine-wave AC. Don't scrimp on this investment!

Figure 4-2 shows the two outlets of my generator. The cord on the left goes to the furnace fan and electronics. The cord on the right goes to my computer

Figure 4-2   My portable generator has two AC outlets. The cord on the left goes to the furnace electronics and fan; the cord on the right goes to the computer workstations.

FIGURE 4-3 Power strip for the cords leading from the generator to the computer workstations. The cords lead to uninterruptible power supplies (UPSs).

workstations, by way of a power strip in the garage. Figure 4-3 shows that power strip, which includes a light bulb that tells me when the generator is running, and also illuminates the garage at night. This power strip *does not* have a transient suppressor (or "surge protector") because the computer workstations both have uninterruptible power supplies (UPSs) with their own transient suppressors.

You should never connect devices with transient suppressors in cascade (one after another) in the same circuit. For example, you shouldn't use a UPS along with a power strip if both devices have transient suppressors. You'll need a power strip without a transient suppressor (they're cheaper that way, anyhow). Transient suppressors in cascade will sometimes interfere with each other's operation, a conflict that can produce bizarre malfunctions! I've seen a UPS "go crazy" with a transient suppressor connected to one of its outputs.

For Nerds Only

Always try to balance the loads among multiple outlets in a generator that has more than one outlet. Ideally, each outlet should do roughly the same amount of work. This precaution ensures that the generator will operate at maximum efficiency. In some generators, the outputs appear in different phases. If I were to connect the entire load to, say, the left-hand outlet in the situation of Fig. 4-2, it would be like seating all the passengers on the left-hand side of an aircraft. The generator would work, but probably not at peak efficiency.

In the event of a utility power failure, I follow a rigorous procedure to disconnect the appliances I want to use from the utility lines before I activate the generator. For example, during a winter storm, the power went out, and I needed to keep the furnace running. I switched off the breaker that controls the furnace, and then followed a step-by-step procedure that I have provided here as Table 4-1. You

TABLE 4-1   Procedure for Generator Use with My Furnace in Case of a Utility Failure

When power goes out:

- Turn OFF the HEAT breaker upstairs.
- Turn OFF the isolation switch on the furnace (move it to the left).
- LOCATE the generator so its exhaust will blow AWAY from the house.
- Be sure the furnace is NOT plugged into the power strip below the isolation switch.
- Be sure NO LOADS are connected to the generator. Switch off ALL power strips.
- START the generator and allow it to run for 3 minutes.
- Be sure the furnace power strip is switched OFF (the light on it should be OFF).
- REMOVE the cap from the furnace emergency plug.
- PLUG the furnace into the power strip.
- Switch the furnace power strip ON (the little light on it should illuminate).
- Switch other power strips back ON if they are to be used.
- Switch ON the generator eco-throttle so it will adjust its speed according to load.
- Set thermostat up to normal temperature.
- KEEP the furnace isolation switch and upstairs HEAT breaker OFF.

When the generator needs refueling:

- DISCONNECT all loads from the generator. Switch ALL power strips OFF.
- Switch the eco-throttle OFF so the generator runs at FULL SPEED.
- Switch the generator OFF.
- Let exhaust pipe COOL before refueling so gas fumes will not ignite.
- Refuel the generator with PREMIUM gasoline, only after the unit has cooled.
- Be extra careful not to overfill the tank.
- START the generator up again and let it run for 3 minutes.
- Switch the furnace power strip ON (the light on it should illuminate).
- Switch other power strips back ON if they are to be used.
- Switch ON the generator eco-throttle so it will adjust its speed according to load.
- KEEP the furnace isolation switch and upstairs HEAT breaker OFF.

When power returns:

- Switch OFF the generator eco-throttle so it runs at FULL SPEED.
- DISCONNECT all loads from the generator. Switch off ALL power strips.
- UNPLUG the furnace from its power strip, once that strip has been turned off.
- REPLACE the cap on the furnace emergency plug.
- Switch OFF the generator.
- Reconfirm that the furnace is UNPLUGGED from the power strip.
- Turn ON the furnace isolation switch (move it to the right).
- Turn ON the HEAT breaker upstairs.
- MOVE the generator back to its sheltered location.

should devise a similar procedure for your own home situation, with the help of an electrician, to ensure that you stay absolutely safe. Write the procedure down in detail, and tape a copy to your furnace. Then, when an outage actually occurs, follow those instructions to the letter.

As a final precaution to keep the generator operating at peak efficiency and safety, you should connect the generator's ground terminal to a known electrical ground, which you have tested for continuity with the main ground for your whole house. Figure 4-4 shows my arrangement, which comprises a single heavy length of wire and a clamp going to a cold water pipe. By performing the ground test described earlier in this book, I've satisfied myself that the cold water pipe connects directly to the main electrical ground for the house.

Warning! Always locate a generator so that its exhaust can vent freely to the outside. The best way to make that happen is to keep the generator outdoors when running it. Never run your generator in a garage (even an open one) or partially enclosed space of any kind. Buy a carbon-monoxide (CO) detector if you don't already have one, and place it in your house near the rooms where you sleep. Keep its batteries fresh. That way, you'll know if generator exhaust "blows" into the house, a situation that can arise with amazing ease, as I discovered when I ran my little Honda generator in the woodshed under my dining room. My CO detector sounded its alarm after only a few minutes of generator time!

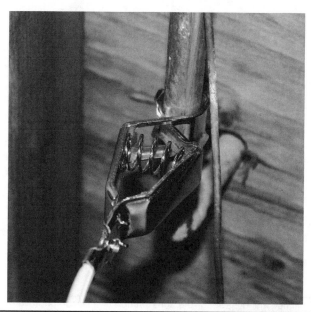

Figure 4-4   Ground clamp for the generator, in this case to a cold water pipe that has been tested to ensure continuity with the main electrical ground for the house.

Warning! An on-site standby generator must run only when your house wiring is completely separated from the electric utility wiring with a *double-pole, double-throw* (DPDT) *isolation switch* installed and tested by a competent, certified electrician. Alternatively, you can plug individual appliances into the generator through dedicated cords that have *nothing whatsoever* to do with your house wiring. If you don't follow these rules strictly, *backfeed* can occur, in which electricity from the generator gets into the utility lines near the home or business where the generator operates. Backfeed can endanger utility workers and damage electrical system components.

Residential Solar Power

A *photovoltaic* (PV) *cell* is a specialized form of *semiconductor diode* that converts visible light rays, infrared (IR) rays, or ultraviolet (UV) rays directly into electricity. When used to obtain electricity from sunlight, this type of device is known as a *solar cell*. One of the most common types, the *silicon PV cell*, is made of specially treated silicon.

Figure 4-5 shows the "innards" of a silicon PV cell. It's made with two types of silicon, called *P type* and *N type*. The functional part is the surface at which these two types of materials come together, known as the *P-N junction*. The top of the assembly is transparent so that rays can strike the junction. The positive electrode

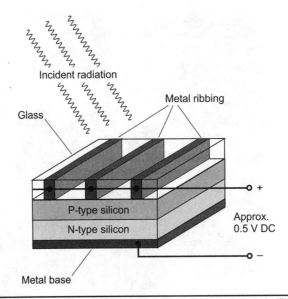

FIGURE 4-5 Functional diagram of the construction of a silicon photovoltaic (PV) cell, also called a solar cell.

is made of metal strips or tiny bars called *ribbing* interconnected by fine wires. The negative electrode comprises a metal base called the *substrate*, placed in contact with the N type silicon. When energy rays (usually in the form of sunlight) strike the P-N junction, a voltage or *potential difference* develops between the P type and the N type materials.

For Nerds Only

The ratio of the available electrical output power to the total radiation power striking a PV cell (with both quantities expressed in the same units, such as watts) is called the *efficiency*, or the *conversion efficiency*, of the cell. The conversion efficiency can, and usually does, vary depending on how much radiation power strikes the cell surface. Any given PV cell will exhibit a different conversion efficiency in a dimly lit room than it will in direct sunlight.

Voltage, Current, and Power

Most silicon solar cells provide about 0.5 V DC with no load connected. If you don't demand that the thing deliver very much current, even moderate light, such as you get on a dreary, overcast day, can generate the full output voltage. As you demand more current, you'll need to have better illumination to produce the full output voltage. An upper limit exists to the current that you can obtain from a particular PV cell, no matter how intense the incident light gets. This limit, called the *maximum deliverable current*, depends on the surface area of the P-N junction, and also on the technology involved in the manufacture of the device.

In a battery consisting of two or more identical PV cells connected in series (negative-to-positive, like the links in a chain), the total voltage increases in proportion to the number of cells, but the maximum deliverable current remains the same as that of any individual cell by itself. In a battery consisting of two or more identical PV cells connected in parallel (negative-to-negative and positive-to-positive, like the rungs in a ladder), the total voltage equals that of any cell alone, but the maximum deliverable current increases in proportion to the number of cells.

When you combine series PV cells in parallel, or parallel PV cells in series, you can get more voltage and more current than you can get from any cell all by itself: the best of both worlds! Engineers call such a set a *series-parallel array* of PV cells.

The *maximum output power* for a silicon PV cell (in watts) equals the product of the output voltage (in volts) and the maximum deliverable current (in amperes). The maximum power that you can get from a series-parallel combination of identical PV cells equals the maximum power from each cell times the total number of cells. When you connect a load to a PV system, and thereby draw current from it, the actual maximum power always turns out slightly lower than the theoretical maximum power.

Did You Know?

If you connect a lot of PV cells in series, you will have to deal with a *voltage drop* of several percent under load as a result of the *internal resistance* of the combination. That's why it's a bad idea to try to get a high voltage by connecting a huge number of PV cells in series. You can do that sort of trick in theory, but in the real world it won't work out very well.

Low, Medium, and High Voltage

In low-voltage, low-current PV systems, individual cells are normally connected in series to obtain the desired output voltage. For charging a 12-V battery, a common PV output level is 16 V, requiring 32 PV cells in series. Such a series-connected set is called a *PV module*. In order to get more maximum deliverable current, multiple modules can be connected in parallel to form a *PV panel*. Finally, if even higher levels are necessary, multiple panels can be connected in series or parallel to obtain a *PV array*.

Although gigantic voltages can theoretically be obtained by connecting hundreds or even thousands of PV cells in series, this approach presents problems because the *internal resistances* of cells in series add up, just as ordinary electrical resistors in series add up. That effect reduces the maximum deliverable current, and it also causes the output voltage to drop under load. High-power PV arrays can be constructed by connecting a large number of cells or low-voltage modules in parallel, making many identical such sets, and then connecting all the parallel sets in series.

If you want to get a medium voltage (say, the nominal voltage for a household utility circuit) from a low-voltage solar panel, you can use a power inverter along with a high-capacity rechargeable battery called a *deep-cycle battery*. The solar panel keeps the battery charged; the battery delivers high current on demand to the power inverter. Such a system provides common 117-V AC electricity from a 12-V DC or 24-V DC source.

Did You Know?

Actual solar panels vary greatly in design and output specifications. Regardless of the voltage level, the maximum available power that you can get in bright, direct sunshine from a solar panel comprising multiple identical PV cells varies in direct proportion to the total surface area of the array.

Mounting and Location

In theory, a solar panel works best when it lies broadside to the incident sunlight, so that the sun's rays shine straight down on the surface. However, this orientation is not critical. Even at a slant of 45 degrees (45°) with respect to the sun's rays, a solar panel receives 71 percent as much energy per unit of surface area as it does when optimally aligned. Misalignment of up to 15° makes almost no difference.

You should locate your solar panels where they will receive as much sunlight as possible, averaged out during the course of the day and the course of the year. Mountings should be sturdy enough so the panels will not rip loose or wiggle out of alignment in strong winds, heavy snow storms, or ice storms. One of the most popular arrangements involves mounting a solar panel, or a set of panels, directly on a steeply pitched roof that faces toward the equator.

The ideal bearing arrangement for a solar panel would be a motor-driven *equatorial mount*, similar to the ones used with astronomical telescopes. This system would allow the panel to follow the sun all day, every day of the year. However, such a sophisticated mechanical device is impractical for most people, and the cost is prohibitive for large panels or multi-panel arrays. The next best thing is a mount with a single bearing that allows for the panel to be manually tilted, always facing generally south in the northern hemisphere or generally north in the southern hemisphere. Figures 4-6 and 4-7 show examples of this type of system.

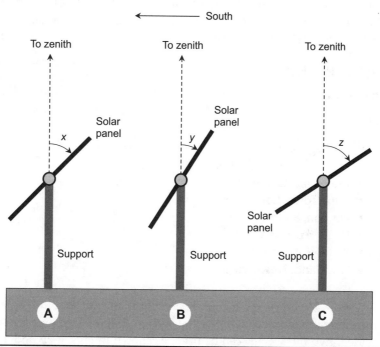

FIGURE 4-6   Optimal placement of fixed, south-facing solar arrays for locations in northern temperate latitudes for year-round operation (A), low-solar-angle-season operation (B), and high-solar-angle-season operation (C). The variables x, y, and z represent angles in degrees with respect to the zenith. In each case the panel is viewed edge-on, looking west.

Northern Hemisphere

The arrangements that you see in Fig. 4-6 will work between approximately 20° north latitude (20° N) and 60° north latitude (60° N). Cities in such locations include:

- Las Vegas, USA
- Chicago, USA
- Miami, USA
- Paris, France
- Berlin, Germany
- Moscow, Russia
- Beijing, China
- Osaka, Japan

Figure 4-6A shows a year-round panel position. You should set the angle x to 90° minus the north latitude at which your system is located. If an adjustable bearing is provided, you can use two tilt settings, as shown in Figs. 4-6B and 4-6C. From late September through late March (autumn and winter), the arrangement shown at B will work the best, and the angle y should be set to approximately 78° minus the latitude. From late March through late September (spring and summer), the arrangement shown at C will work the best, and the angle z should be set to approximately 102° minus the latitude.

Southern Hemisphere

The arrangements in Fig. 4-7 will work between approximately 20° south latitude (20° S) and 60° south latitude (60° S). Cities in such locations include:

- Santiago, Chile
- Buenos Aires, Argentina
- Rio de Janeiro, Brazil
- Cape Town, South Africa
- Durban, South Africa
- Perth, Australia
- Sydney, Australia
- Auckland, New Zealand

Figure 4-7A shows a year-round panel position. You should set the angle x to 90° minus the south latitude at which your system is located. If an adjustable bearing is provided, two tilt settings can be used, as shown in Figs. 4-7B and 4-7C. From late March through late September (autumn and winter), the arrangement shown at B is optimal, and the angle y should be set to approximately 78° minus the latitude. From late September through late March (spring and summer), the arrangement shown at C is optimal, and the angle z should be set to approximately 102° minus the latitude.

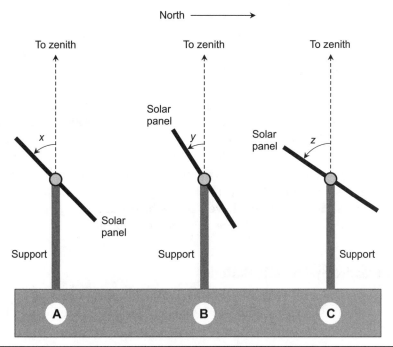

Figure 4-7   Optimal placement of fixed, north-facing solar arrays for locations in southern temperate latitudes for year-round operation (A), low-solar-angle-season operation (B), and high-solar-angle-season operation (C). The variables *x*, *y*, and *z* represent angles in degrees with respect to the zenith. In each case the panel is viewed edge-on, looking west.

Stop and Think!

Before you invest thousands of dollars in a solar power system for your house, consult a competent engineer who can assess your situation. Here are some things to think about.

- You can't expect a photovoltaic system to provide as much power as your electric utility company does.
- Photovoltaics only provide power to a system when the sun shines brightly enough. Small-scale PV systems rarely justify the cost in locations that don't receive much sunlight.
- If the solar panels get covered with snow or debris, you'll have to manually remove the obstruction if you want the system to keep working.
- Problems with load imbalance can occur if part of a solar array lies in bright sunlight while another part lies in shadow. You'll have to find a location that gets plenty of continuous, total sun exposure during much of the day.
- In a PV system that uses lead-acid storage batteries, the batteries can produce dangerous fumes. All types of rechargeable batteries require maintenance, and you'll have to replace them altogether every few years. That can be quite expensive for a large solar power plant.

Stand-Alone System

A *stand-alone small-scale PV system* uses rechargeable batteries to store the electric energy supplied by a PV panel or array. The batteries provide power to an inverter that produces 117 V AC (in the United States). In some systems, the battery power can be used directly, but this method will work only with home appliances designed for low-voltage DC. Figure 4-8 is a functional block diagram of a stand-alone small-scale PV system that can provide 117 V AC for the operation of small appliances in a typical American household.

The batteries allow the system to produce usable electricity even if there's not enough sunlight for the PV cells to operate, for as long as the batteries retain some charge. A stand-alone PV system of this type offers independence from the utility companies. However, you'll have a power blackout if the system goes down for so long that the batteries discharge and you don't have a backup power source, such as a generator.

Interactive System with Batteries

An *interactive small-scale PV system with batteries* resembles a stand-alone system, but with one significant addition. If you get a prolonged spell without enough light

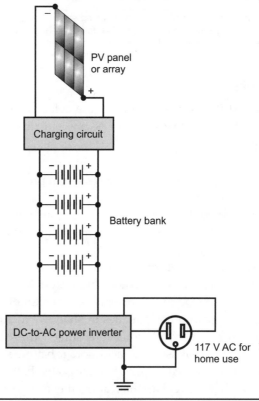

FIGURE 4-8   A stand-alone small-scale PV system.

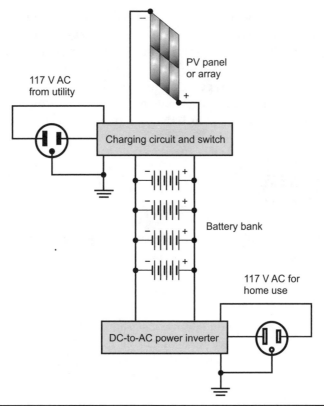

FIGURE 4-9   An interactive small-scale PV system with batteries.

for the PV cells to function, the electric utility can take over, keeping the batteries charged and preventing a blackout. A switch, along with a battery-charge detection circuit, connects the batteries to the utility through a charger if insufficient power, or no power at all, comes from the PV panel or array. When conditions become favorable and the PV cells can work again, the switch disconnects the batteries from the utility charger and reconnects them to the PV panel or array.

Figure 4-9 is a functional block diagram of an interactive small-scale PV power system with batteries. In this arrangement, you don't sell any power to the electric utility company, even when the PV panel or array generates more power than your home needs. When the utility is involved, the electrical energy only flows one way, from the utility line to the batteries through a charging circuit and switch. That situation occurs only when the batteries require charging and the PV panel or array does not provide enough power to charge them.

Interactive System without Batteries

An *interactive small-scale PV system without batteries* operates in conjunction with the utility company, just as the system with batteries does. You can sell energy to the company during times of minimum demand, and buy it from the company at times of heavy demand. With this type of system, you can keep using electricity (by buying it directly from the utility) if there's a long period of dark weather, and you don't have to care for a set of batteries. Another advantage is that, because no batteries are used, this type of system can have greater peak-power-delivering capability than a stand-alone arrangement or an interactive system with batteries. Figure 4-10 is a functional block diagram of an interactive small-scale PV system without batteries.

Did You Know?

In the United States, some states offer good buyback deals with the utility companies, and some states don't. Check the utility buyback laws in your state before you decide to invest in an interactive electric generating system of any kind.

Direct PV Climate Control

Figure 4-11 is a simplified block diagram of a direct PV system for indoor environment modification. Remember that in bright sunshine, a single silicon PV cell produces approximately 0.5 V DC. You can connect numerous silicon PV cells in a *series-parallel array* that provides 12 V DC or 24 V DC output at fairly high current in direct sunlight.

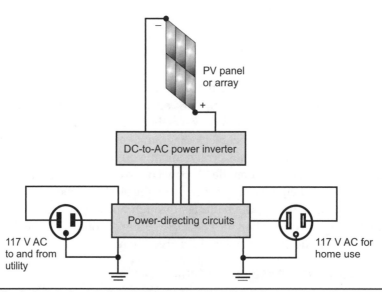

Figure 4-10   An interactive small-scale PV system without batteries.

Figure 4-11 shows only four PV cells (for simplicity), but it illustrates the basic series-parallel principle. A real-world system can contain hundreds of individual PV cells. A *solar module* of 53 silicon PV cells connected in series, each rated at 0.5 V DC, theoretically yields 26.5 V DC with the same maximum current output as a single cell. When you call upon the system to produce power, this figure drops to around 24 V DC because of the internal resistance of the PV cells. By connecting multiple 53-cell series modules in parallel to form a *solar panel*, you can obtain high current levels at the same voltage (in this case 24 V DC).

The output of the solar panel goes to a power inverter that changes the low-voltage DC output of the solar panel into 117 V AC that can operate electric heaters. The system includes a voltage regulator to ensure that the voltage remains fairly constant under conditions of varying solar intensity. This type of system needs a high-current power inverter, and that thing can cost a lot of money. But you do have an alternative, if all you want to do is run a small electric heater from your PV

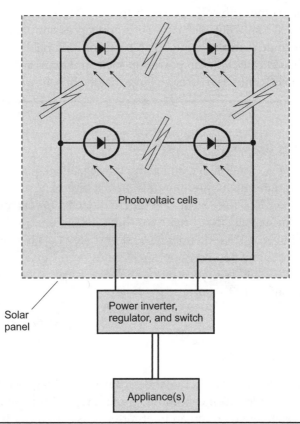

FIGURE 4-11   A supplemental indoor environment control system that uses a solar panel, power inverter, voltage regulator, and switch connected to a small appliance, such as a fan or humidifier.

array. You can do away with the inverter altogether, design your PV array to produce 117 V DC, and then supply the heater with 117 V DC instead of 117 V AC, remembering that you can't run most other household appliances from DC. You'll also have to keep in mind the fact that you can't expect to run a whole household full of electric heating elements with a system like this (unless you have the money and real estate to build a massive "solar ranch").

A well-designed direct PV system has an automatic shutdown switch that disconnects the solar panel if the daylight becomes too dim to properly operate appliances connected to it. If the system runs near peak capacity and the delivered current suddenly drops (a storm cloud moves in, for example), the switch will power down the system until sufficient daylight returns.

Fact or Myth?

- People have said that you can't run electric baseboard heating systems, central air conditioners, or lots of heavy appliances (in general) all at the same time with solar power. Is that true?
- If you run a lot of heavy-duty appliances simultaneously, they'll draw too much current for a solar panel of reasonable size to contend with. Theoretically, you can power up such appliances with solar panels, but the panels would have to be so large that the benefit would not justify the cost.

For Nerds Only

Electric fans, humidifiers, and evaporative coolers don't draw much current, so you can use them intermittently with a direct PV source of power. During the summer in a place such as southern Arizona, for example, the hottest part of the day usually has bright sunshine. In a scenario like that, the system shown in Fig. 4-11 could operate a set of ceiling fans in a home or business. In a cold but sunny desert region, such as northern Nevada or central Wyoming, the system could operate a humidifier to mitigate the extreme low-humidity indoor conditions that prevail in winter.

Residential Wind Power

The term *small-scale* applies to wind turbines that can generate up to about 20 kW of electricity under ideal conditions, enough to power most households. All wind turbines generate power on an intermittent basis. In order to obtain a continuous supply of electricity with a small-scale wind-power plant, you'll have to use storage batteries or an interconnection to the electric utility, or both.

Stop and Think!

Before you decide to install a wind turbine for your house, think about the following potential pitfalls. As with a solar electric system, you should consult a reputable engineer who specializes in alternative energy.

- Some places have a lot more wind than others. Ask yourself how your locale "rates" in this department, and answer yourself honestly! You can find wind maps on the Internet. Their addresses keep changing all the time, so you'll probably want to enter a phrase like "wind power map" into your favorite search engine.
- Even in the windiest places, such as Wyoming or South Dakota or Nebraska, the wind doesn't blow all the time.
- Small-scale wind turbines will not work properly if the wind gets too strong.
- A small wind turbine can be wrecked by a powerful thunderstorm, hurricane, or ice storm.
- It will take a long time to recoup the up-front installation cost, even if you are using a small-scale wind-power system.
- Your neighbors may dislike having a wind turbine nearby.
- Small-scale wind turbines can create significant noise at close range.

How It Works

Most small-scale wind turbines are steered by a *wind vane* attached to the generator housing (called the *nacelle*). The vane works in the same way that an old-fashioned weather vane does. When the wind blows hard enough to operate the turbine, the vane orients itself to point away from the wind. Under normal operating conditions, the plane defined by the blade rotation lies perpendicular (broadside) to the wind direction.

In a small-scale wind-power system, the speed of the blade rotation vares with the wind speed, resulting in variable-frequency AC from the generator inside the nacelle. This generator resembles the *alternator* in a motor vehicle. (Some manufacturers call it an alternator for that reason.) The AC from the generator is converted to DC by a rectifier circuit, and the DC charges a set of storage batteries. The electricity for household appliances comes from these batteries either directly, in which case special DC appliances must be used, or by means of a power inverter that converts the low-voltage DC electricity from the batteries to 117 V AC at 60 Hz (in the United States) or 50 Hz (in Europe and some other parts of the world).

The plane defined by the blades is normally perpendicular to the axis of the vane, so that the wind blows straight at the blades. However, in a strong wind, the plane of the blades changes, so it no longer lies perpendicular to the vane axis. This adjustment reduces the wind load on the blades but allows the turbine to keep on working. As the wind speed grows stronger yet, the angle between the plane of the blades and the vane axis decreases until, at a certain speed, it becomes zero. Then the blades rotate in a plane that contains the axis of wind flow. The variation in the angle between the plane of the blades and the wind direction is called *furling*. It can be done in the horizontal plane (so the blades swing, or *yaw*, toward the left or right) or in the vertical plane (so the blades tilt up or down).

A wind turbine can also regulate its wind load by varying the *blade pitch*. When the blade pitch is small (the plane of each blade's surface is nearly the same as the plane defined by the blades), the wind produces less torque in the system, and consequently less power, than when the blade pitch is large (the plane of each blade's surface differs greatly from the plane defined by the blades). At low wind speeds, the blade pitch is at the maximum. As the wind speed increases, the blade pitch decreases. If the wind speed becomes great enough, the blade pitch becomes zero.

Did You Know?

In extremely high winds, the blades can turn to zero pitch, furl completely, and lock in place. This maneuver reduces the load on the blades as much as possible, minimizing the risk of structural damage. It also shuts down the turbine.

Stand-Alone System

A *stand-alone small-scale wind-power system* uses rechargeable batteries to store the electric energy supplied by the rectified output of the generator. The batteries provide power to an inverter that produces a "clean" AC wave at 117 V. The very best inverters produce "true sine waves." The second-best ones produce "modified sine waves."

Some stand-alone systems use the battery power directly without any inverter at all, but this arrangement will work only with appliances and devices designed to run from low-voltage DC. Figure 4-12 is a functional block diagram of a stand-alone small-scale wind-power system that can provide 117 V AC.

The use of batteries allows the system to produce usable power even if there's not enough, or too much, wind for the turbine to operate. A stand-alone system offers independence from the utility company. However, a blackout will occur if the system goes down for so long that the batteries discharge and no backup power source exists.

Interactive System with Batteries

An *interactive small-scale wind-power system with batteries* resembles a stand-alone system, but with one significant addition. If you suffer through a prolonged spell

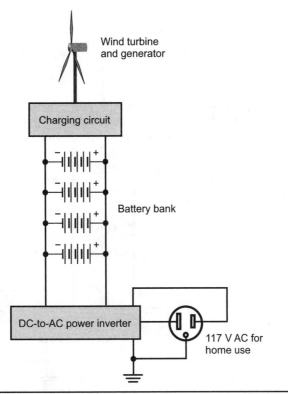

FIGURE 4-12   A stand-alone small-scale wind-electric system.

in which wind conditions are unfavorable for turbine operation, the electric utility can take over to keep the batteries charged and prevent a blackout. A switch, along with a battery-charge detection circuit, connects the batteries to the utility through a charger if no power issues from the turbine. When wind conditions become favorable and the turbine supplies power again, the switch disconnects the batteries from the utility charger and reconnects them to the turbine generator and rectifier.

Most interactive small-scale wind-power systems with batteries never sell any power to the electric utility, even if the wind turbine generates an excess. Power only flows one way, from the electric power line to the batteries through a charging circuit and switch, and even that happens only when the batteries require charging and the wind turbine does not provide enough power to charge them. Figure 4-13 is a functional block diagram of this type of wind-power system.

Interactive System without Batteries

An *interactive small-scale wind-power system without batteries* also operates in conjunction with the utility company. You sell excess energy to the company during times of

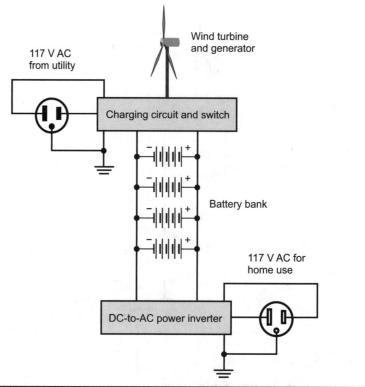

FIGURE 4-13   An interactive small-scale wind-electric system with batteries.

minimum demand, and buy energy from the company during times of heavy demand. You can keep using electricity (by buying it directly from the utilities) if wind conditions remain unfavorable for a prolonged period. Because this type of system has no batteries, it can be larger, in terms of peak power-delivering capability, than a stand-alone arrangement or an interactive system with batteries.

An interactive system without batteries, like the type with batteries, is designed to function with the help of the utility company, and does not offer the independence that a purist might desire. This factor does not represent a technical drawback, but it can pose a philosophical problem for anyone who desires to live completely off the grid. Figure 4-14 is a functional block diagram of an interactive small-scale wind-power system without batteries.

Direct Wind-Powered Climate Control

Figure 4-15 illustrates a wind turbine, equipped with an electric generator and connected into a zone electric baseboard heating system. A voltage-regulation circuit maintains the system at or near 117 V AC, so the heating elements can operate as they normally would with the electric utility.

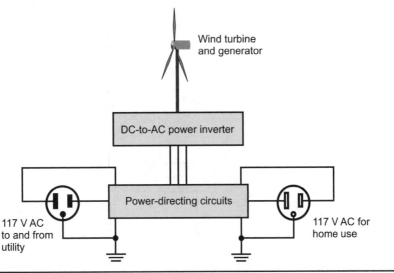

Figure 4-14   An interactive small-scale wind-electric system without batteries.

A medium-sized wind turbine designed for residential use can produce about 12 kW of power on a day with moderate wind. That's the equivalent of eight electric space heaters, each rated at 1500 W. As things work out, 1 kW of electrical power equals 3410 British thermal units per hour (Btu/h) of "heating power." Therefore, the wind turbine system of Fig. 4-15 can provide approximately 3410 × 12 = 40,920 Btu/h. A gas furnace for a typical residential home produces 80,000 to 100,000 Btu/h when running full blast. So in theory, the system shown in Fig. 4-15 can supply about half of the energy necessary to keep your house warm.

A system like the one shown in Fig. 4-15 depends on wind for its operation. Batteries of reasonable cost can't store the large amounts of energy required for home heating. In a location where the wind does not blow often or hard enough, this scheme won't prove cost effective. However, in some places, winters remain cold and windy for weeks or months at a time. Such places make good "proving grounds" for a system such as the one diagrammed in Fig. 4-15.

Did You Know?

With a conventional wind turbine, the available power from a wind of constant speed varies in direct proportion to the square of the blade length (that is, the square of the turning radius of the whole blade system). That's because the area of a circle (the shape of the region swept out by the blades as they rotate) is proportional to the square of its radius. This rule assumes that you don't change the number of blades or the general shape and pitch of each blade.

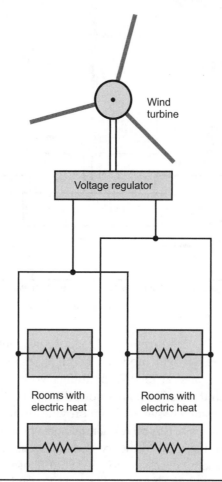

FIGURE 4-15 A supplemental residential heating system that uses a wind turbine and voltage regulator connected into a conventional electric zone heating circuit.

For Nerds Only

If the wind speed doubles, say from 10 mi/h to 20 mi/h, then the theoretical amount of available wind power (the power you can get from the wind with a turbine or similar device) increases by a factor of 8 (or 2^3). If the wind speed triples, say from 10 mi/h to 30 mi/h, then the theoretical wind power becomes 27 (or 3^3) times as great. If the wind speed quadruples, say from 10 mi/h to 40 mi/h, then the theoretical wind power becomes 64 (or 4^3) times as great!

Residential Hydro Power

A water turbine designed for a home or small business, installed in a fast-moving stream or small river with sufficient vertical drop, can produce roughly 20 kW of electricity, more than enough for a typical household under conditions of peak demand.

Stop and Think!

Residential hydro power systems won't work for as many people as small-scale wind or solar power systems will, and for good reasons. If you've contemplated a hydro system for your own home, you should consider the following factors before you proceed.

- Only a few people live on properties with streams that provide enough flow to provide hydroelectric power. Be honest with yourself: Are you among them?
- You'll have to verify your water rights before you modify the water resources on your property, so that you know what you can legally do (or not do).
- A small stream might periodically completely freeze or dry up, shutting a small-scale hydropower system down. How cold does it get in the winter where you live? Have you ever checked out your stream at the nadir of the winter season to see if any water flows?
- A water turbine requires considerable water mass, along with a significant vertical drop, to provide enough power to effectively serve a residence. You might have to install a small dam or artificial waterfall to build a workable system, and these arrangements could give rise to environmental and regulatory issues.
- The up-front cost of a small-scale hydropower system is considerable. It takes a long time to pay for itself, and the resulting economic benefit may be outstripped by the initial cost.

How It Works

A small-scale hydropower system can be configured in three ways: stand-alone, interactive with batteries, and interactive without batteries. These three types of systems work in the same way for small-scale hydro-power systems as they do for small-scale residential wind-power systems.

Most small-scale hydroelectric systems use *diversion technology*, in which a portion of a river or fast-moving stream is channeled through a canal or pipeline, and the current through this medium drives a water turbine. You don't need a dam. This type of system works best in locations where a river drops considerably per unit horizontal distance. Small-scale and medium-scale diversion systems can be used next to mountain streams or fast-moving, small rivers for the purpose of providing energy to homes.

An *impoundment hydroelectric power plant* consists of a dam and reservoir. This type of facility works best in mountainous places where high dams can be built and deep reservoirs can be maintained. Figure 4-16 is a simplified functional diagram of an impoundment facility. The water from the reservoir passes through a large pipe called a *penstock*, and then through one or more water turbines that drive one or more electric generators.

A *pumped-storage hydroelectric power plant* has two or more reservoirs at different elevations. When there's little demand for electricity, the excess available power is used to pump water from the lower reservoir into the upper one(s). When demand increases, the potential energy stored in the upper reservoir(s) is released. Water flows out of the upper reservoir(s) in a controlled manner, passing through penstocks and turbines to generate electricity.

Did You Know?

Pumped-storage systems require dams to hold the water in the reservoirs. These dams are generally smaller than the dams used in large impoundment facilities. Pumped-storage power plants can be found in regions where the terrain is hilly or gently rolling. If you're lucky enough to have a large ranch in a mountainous region, this type of system might work well for you. But a significant difference in average elevation must exist between the reservoirs.

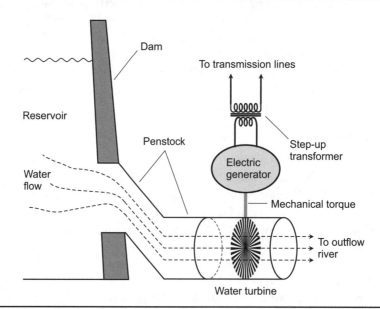

Figure 4-16   A hydro power system that derives its energy from water impoundment.

Fact or Myth?

Suppose that you'd like to install a stand-alone, small-scale, impoundment-type hydro power system for your ranch, but you're concerned about the effect it will have on wildlife. Imagine that a fairly good-sized stream runs through your property, that an engineer has checked everything out, that the vertical drop is sufficient, and that there's more than enough water flow all year round. You'll need to build a small dam and back up some water to form a pond (or even a small lake). You've checked everything out with the local, state, and federal officials, and they're okay with your plans. Naturalists from a nearby college or university can offer some insight as to what effects your system will have on wildlife (and you can rest assured that some effects will occur). A pond can be expected to attract birds, fish, and other wildlife. It might even serve as a watering spot for your cattle! However, the same pond will displace other wildlife, particularly mammals that dwell beneath the surface. Are all small-scale stand-alone hydro power systems bad for the planet? No. Will yours harm the environment in general? You'll have to figure that out for yourself, with the help of objective advisors.

Direct Hydroelectric Climate Control

You can connect a water turbine to an electric generator, which can drive electric heating and cooling systems in much the same way as a wind turbine can do. With sufficient water flow and proper voltage regulation, an arrangement of this kind can provide some of the power for climate control in a typical household.

Figure 4-17 is a block diagram of a small water-driven energy system adapted for use with electric baseboard heating. This assembly resembles the direct wind-powered system, except that you'll replace the wind turbine with a water turbine. As with the wind system, a regulator circuit keeps the voltage near 117 V AC.

An efficient water turbine, installed in a fast-moving stream or small river with sufficient vertical drop, can produce 20 kW of power on a reliable basis. Again, recall that 1 kW = 3410 Btu/h. Therefore, a substantial water turbine system can provide approximately 3410 × 20 = 68,200 Btu/h. (Let's round this off to 70,000 Btu/h). That amount of power can keep a small home comfortable in almost all types of weather, as long as the stream or river doesn't dry up or freeze solid.

Quick Question, Quick Answer

- You ask, "A stream runs through my property. It's 10 meters (33 feet) wide and 2 meters (6 feet) deep in the middle. It flows well except in winter, when it freezes on the surface, although the water keeps flowing under the ice. The vertical drop, from the point where the stream enters

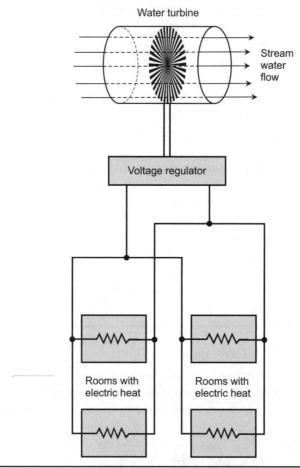

FIGURE 4-17   A supplemental residential heating system that uses a water turbine and voltage regulator connected into a conventional electric zone heating circuit.

the property to the point where it exits, amounts to only 0.5 meters (approximately 18 inches). Will this stream provide enough hydroelectric power to heat my home?"

- You should get an engineer to evaluate your situation whenever you contemplate the installation of a residential hydro power system. Based on the foregoing information, your stream probably won't provide enough power to heat your home. A water turbine in this stream could produce a few watts of electricity on a reliable basis, enough to operate a few small lamps and keep a notebook computer charged.

Deep-Cycle Batteries

You'll need *deep-cycle batteries* to operate any type of stand-alone alternative-electric system. Deep-cycle batteries are usually of the lead-acid type, and in that respect, they resemble automotive batteries. But there are a couple of big differences between automotive batteries and deep-cycle batteries.

Deep-Cycle versus Automotive

Automotive batteries can provide lots of current for a very short time; you need that kind of current to start your car or truck. A typical automotive battery can produce around 750 A of *cold-cranking current*. That's the current the vehicle demands from the battery when you start it up after it hasn't run for a while. At 12.6 V (the typical voltage of an automotive battery), 750 A give you a little less than 10 kW, roughly the amount of power that your house will likely need with most of your appliances running at once! Obviously that battery, contained in a chamber about the size of a bread box, can't deliver 10 kW for very long. But that current "surge" will be there when you need it, as long as you keep the battery charged up. A deep-cycle battery, in contrast, is not designed to produce anywhere near that much current, even for a brief moment. A deep-cycle battery is intended to produce moderate current for extended periods of time.

Did You Know?

A deep-cycle battery *might* work as a starting battery for your vehicle, especially if you have a car or truck with an electronic ignition system (most do these days). But you'll have to oversize the battery, and that means you'll have to spend more money than you would spend on a conventional automotive battery.

Here's the other difference between automotive and deep-cycle batteries: Automotive batteries are meant to be kept at a full charge, or nearly a full charge, all the time. They're not intended to provide power all by themselves for hours on end. You don't need to discharge your automotive battery while you're driving down the highway; the vehicle alternator keeps it charged up. Have you ever found out what happens when your vehicle's alternator fails? The vehicle will run okay as you drive between two towns several hours apart, but when you stop and switch off the engine, you won't be able to start it up again without a "jump" from another vehicle. On the other hand, a deep-cycle battery is designed to produce current for a long time on its own, losing much of its charge in the process. Most deep-cycle batteries work best if you let them discharge about halfway with each *charge-discharge cycle*. Sometimes you can let a deep-cycle battery discharge 75% or 80% of the way down, but you should never let it lose *all* of its charge. Solar- and wind-powered stand-alone systems need a battery that can "carry the load" for extended periods. A deep-cycle battery can do a good job of that. An automotive battery can't.

For Jocks Only

You might imagine an automotive battery as the equivalent of a "sprinter" who performs well in the 100-meter track event or the 50-meter swimming event. Thinking along the same lines, then, a deep-cycle battery is like an "endurance athlete," such as a runner who excels in the 10,000-meter event, or a swimmer who specializes in the 400-meter and 1,500-meter events.

Did You Know?

Deep-cycle batteries, like automotive batteries, work best if you charge them up slowly. Engineer call that process *trickle charging*. In any well-designed alternative-electric system with batteries, a *charge controller* makes sure that the battery bank receives the optimum amount of charging current at all times. You should avoid *quick charging* because it doesn't give you the best battery performance, and in fact, the excessive current can damage your batteries.

How Long Do They Last?

You can expect a deep-cycle battery to last several years, although some batteries perform better than others in this respect. The typical deep-cycle battery might wear out after five to seven years; the very best ones (Crown and Rolls, for example) can last upwards of 15 years. No matter what the quality level of your batteries (and no matter how much money you spend on them), you'll have to treat them properly if you want them to endure for their rated life spans. Here are some recommendations.

- The further "down the cycle" you discharge your batteries, on the average, the sooner they'll wear out.
- Ideally, you should not let your batteries discharge below the 50% level on any cycle; or, if you have to do it, you should not do it very often.
- If possible, avoid allowing your batteries to discharge below the 20% level at any time.
- Always use a charge controller with your batteries. This precaution will prevent overcharging, which can ruin a set of batteries in a hurry. See the section about charge controllers below.
- Have a professional choose the optimum storage capacity, in ampere-hours or watt-hours, for your battery bank. That capacity will depend on the size

of your PV array, wind turbine, or water turbine. It will also depend on how much electricity you expect to get from your system during the course of your everyday living.

- Try to keep your batteries at or near room temperature. Avoid letting them sit in an environment where the temperature drops below freezing.
- Battery capacity goes down as the temperature goes down, even at temperatures above freezing.
- Never try to quick-charge your batteries. Always trickle-charge (slow-charge) them.
- When you buy a set of batteries, start using them right away. You can't store batteries for a long time, and then expect them to last as long as they should when you finally get around to using them.
- When you combine batteries in series or parallel, make sure that all the batteries are identical. Don't combine batteries with different ampere-hour capacities or different voltages.
- When you clean the outside casing of a battery, use only distilled water. Don't use any of those high-tech cleaning concoctions.
- Read, heed, and save all of the instructions provided with your batteries when you buy them! Pay special attention to the directions for adding fluid to batteries that need to be periodically "watered."

Quick Question, Quick Answer

- In the olden days, instruction manuals for cell phones and laptop computers said that you should discharge the batteries almost all the way down before recharging them again. Does that principle hold true for today's deep-cycle batteries intended for stand-alone alternative-electric systems?
- Absolutely not. Deep-cycle batteries aren't like the old nickel-cadmium (NICAD) types that were meant to be discharged all the way down. Today's deep-cycle batteries *do not* suffer from the *memory drain* problems that made NICADs notorious. If you repeatedly discharge a deep-cycle battery bank almost all the way down, you'll probably shorten its life.

Charge Controllers

If you plan to charge deep-cycle batteries with the intent of using them in an alternative-electric system of any kind, you'll need a charge controller to limit the rate at which the current goes into, or comes out of, your batteries. The charge controller prevents the batteries from overcharging or getting too charge-depleted. Either of those conditions will shorten the working life of your battery bank considerably, and might even wreck it straightaway. Charge controllers come in two basic types.

1. A *series charge controller* keeps your batteries from overcharging by shutting down the charging current (disconnecting it) when the batteries reach a state of 100% charge.
2. A *shunt charge controller* diverts the current from the PV panel, wind turbine, or water turbine to an auxiliary load, such as a set of electric lights, when the batteries have reached full charge.

The best charge controllers have meters that will tell you how much current you're using at any given time, and how much voltage your batteries are producing at that time. You should consult a professional who will recommend the best type of charge controller for your system when you select or build it.

Fuel Cells

In the late part of the twentieth century, a new type of electrochemical power device emerged that holds promise as an alternative energy source: the *fuel cell*. This device converts combustible gaseous or liquid fuel into usable electricity, but at a lower temperature than normal combustion does. In practice, a fuel cell behaves like a battery that you can recharge by filling a fuel tank, or if the fuel is piped in, by a continuous external supply.

What's a Fuel Cell, Anyway?

The most talked-about fuel cell during the early years of research and development became known as the *hydrogen fuel cell*. As its name implies, it derives electricity from hydrogen. The hydrogen combines with oxygen (it *oxidizes*) to form energy and water, along with a small amount of nitrous oxide if air serves as the oxidizer. When a hydrogen fuel cell "runs out of juice," a new supply of hydrogen will get it working again.

Instead of literally burning, the hydrogen in a fuel cell oxidizes in a controlled fashion, and at a much lower temperature. Several schemes exist for making this process go smoothly. The *proton exchange membrane* (PEM) *fuel cell* represents one of the most widely used technologies. A PEM hydrogen fuel cell generates approximately 0.7 V DC, a little less than half the voltage of a typical electrochemical dry cell. To get higher voltages, individual cells are connected in series, so that their voltages add up. For example, to obtain 14 V DC, we would connect 20 hydrogen fuel cells in series because 20×0.7 V = 14 V. A series-connected set of fuel cells technically forms a battery, but engineers and technicians more often use the term *stack*.

Increased current-delivering capacity can be obtained by connecting cells or stacks in parallel, so that the current-delivering capacities of the individual cells or stacks add up. (The voltage of a parallel-connected set of identical cells or stacks equals the voltage of any single cell or stack all by itself.) For example, if you connect five stacks in parallel, each rated at 14 V DC and capable of delivering up to 10 A, the resulting combination will provide 14 V DC at up to 50 A because $5 \times 10 \text{ A} = 50 \text{ A}$.

Fuel-cell stacks can be obtained in various sizes from commercial vendors. A stack about the size and weight of a suitcase full of paperbound books can power a subcompact electric car. Smaller cells, called *micro fuel cells*, can provide electricity for portable radios, lanterns, notebook computers, and other devices that have historically operated from conventional cells and batteries.

A fuel cell can get its "juice" from energy sources other than hydrogen. Almost any liquid or gas that will combine with oxygen to generate energy has aroused interest among engineers. *Methanol*, a form of alcohol, is easier to transport and store than hydrogen because it exists as a liquid at room temperature. Propane and methane have been used to provide the energy for fuel cells. Even gasoline, petroleum diesel fuel, and biodiesel fuel can do the job!

How It Works

Figure 4-18 is a functional block diagram of a small-scale fuel-cell power plant suitable for a home or small business. This system can also work for recreational vehicles (RVs) and boats. In the case of a fixed land location, the fuel can be stored on site or piped in. Engineers have suggested conventional methane as an ideal fuel source for home power plants of this type because the delivery infrastructure exists right now, and on-site storage is not necessary. However, in rural areas, or in any location not served by methane pipelines, other fuels might prove to be more cost-effective.

A typical fuel-cell stack delivers several volts DC, comparable to the voltage produced by a solar array or automotive battery. Under normal conditions, the DC from the fuel cell goes to a power inverter that produces usable 117 V AC output from the low-voltage DC input. If desired, a backup battery bank can keep the electric current flowing when the fuel tank is refilled. A power control system switches the electrical appliances between the fuel cell and the battery bank as necessary.

Did You Know?

The DC output from the fuel cell or battery bank can directly provide electrical power to small appliances designed to run on low-voltage DC, such as two-way radios, some hi-fi sound systems, and notebook computers.

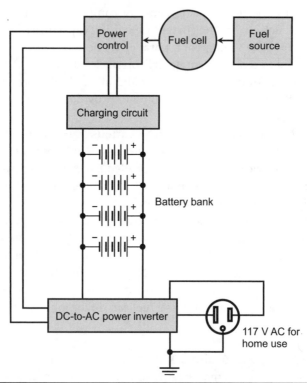

Figure 4-18   A small-scale fuel-cell-based electric power plant.

Quick Question, Quick Answer

- You might reasonably ask, "Can I, as a homeowner, expand a system, such as the one shown in Fig. 4-18, to take advantage of other sources of energy, such as sunshine or wind? If so, how might I do it? Could such a scheme allow my home to operate entirely off the electric utility grid?"
- You can do it, but you'll probably have to spend a lot of money to make it happen. For example, a solar panel or array can charge the battery bank during sunny weather. A wind turbine can supplement the solar source, taking over on windy nights or windy, overcast days. The fuel cell can operate when neither wind nor solar energy can meet your electrical needs. A computer-governed power-control switch can ensure that the system uses the available energy in the most efficient manner at all times. Such a hybrid system could, in theory, offer complete independence from the electric utility. The key to success in such an endeavor would lie in the diversity and redundancy of your energy sources. It wouldn't hurt if you're wealthy, either.

CHAPTER 5

Electronics in Your Vehicle

For most people, the electronics in a car or truck comprises a computer along with various gadgets and "black boxes." If something goes wrong with any of those things, you get your vehicle to a shop for repair, the technician fixes the problem, and that's that. But you can maintain or repair some of the electronics yourself in even the most advanced motor vehicle. In this chapter, we'll focus on things that you, the everyday motorist, have some control over: the battery, the charging system, the fuse panel, the lights, the two-way radio, and the hi-fi system.

The Battery

In any car, truck, or other vehicle with an electrical ignition system, a battery provides the current surge that's necessary to get the engine going. The engine can keep running without a battery once it's started, but you can't start it without a good battery. It's not like a simple lawn mower, snow blower, or small gasoline generator where you just tug on a rope.

How It Works

A typical automotive battery is a heavy box about half the size of a concrete cinder block. The battery gets its heft from the elemental lead that it contains. In the United States, a standard automotive battery produces 12.6 V at full charge with no load connected (that is, when nothing demands any current from it). It contains six 2.1-V lead-acid cells connected in series. A fully charged battery in good condition can generate a great deal of current for short periods of time when called upon to do so, typically several hundred amperes. Figure 5-1 is a simplified functional diagram of a lead-acid cell of the sort your vehicle's battery contains.

When a lead-acid cell has a full charge, the negative electrode comprises essentially pure lead, and the positive electrode comprises *lead dioxide*, a compound of lead and oxygen. The electrolyte, which contains all the battery's energy in

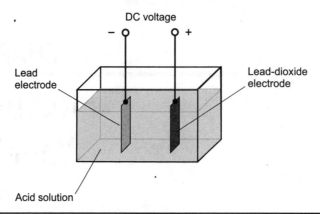

FIGURE 5-1   Simplified functional diagram of a lead-acid cell in its fully charged state.

chemical form (and that converts to electrical energy when something demands current), is a solution of roughly one part sulfuric acid to two parts water. The cells produce their "juice" as a result of a chemical reaction between the electrodes and the acid solution. As the battery discharges, four major things happen.

1.   The negative electrode gradually changes from pure lead to *lead sulfate*, a compound consisting of lead, sulfur, and oxygen.
2.   The positive electrode gradually changes from lead dioxide to lead sulfate.
3.   The acid solution gradually grows more dilute (less sulfuric acid and more water).
4.   Electrons are liberated from the atoms in the solution (that's the electricity).

If you use the battery for a long time without charging it, the voltage gradually decreases, so that when the battery has "run out of juice," it will produce about 11.7 V with no load. That might not seem like much of a drop from 12.6 V at full charge, but the battery's ability to produce current goes down almost to zero. That's why, if your car or truck has a "dead battery," you might manage to run your radio and the interior lights for a little while; but when you turn on the headlights they'll shine dimly, if at all, and you won't be able to start the engine.

Warning!   Even though the sulfuric acid in an automotive battery is diluted with water, the solution can do a lot of damage, and it can cause serious injury if it gets on your skin. You'll get a chance to realize this fact if you happen to come into contact with it. (Don't!) I've seen "battery acid" eat a penny-sized hole in a pair of nylon running pants within seconds. Trust me, I know. I had the pants on when it happened. (And yes, the damage occurred in the worst possible place.)

Why It Fails

An automotive battery won't work if the weather gets so cold that the solution freezes inside it. In some parts of the United States, the temperature in the middle of the winter can drop to minus 40°F (minus 40°C) at times, and once in awhile it gets even colder than that in certain places like Montana, Wyoming, the Dakotas, Minnesota, Wisconsin, Maine, and Alaska. If you leave your vehicle outdoors all night when it's that cold, you shouldn't be surprised if you find a "dead battery" in the morning.

Did You Know?

If your vehicle has an engine-block heater that you can plug into an electrical outlet, you should use it in exceptionally cold weather; but it won't do any good for your battery. It'll keep your engine oil from congealing, but it won't keep your battery electrolyte from freezing.

All automotive batteries eventually lose their ability to hold a charge because they literally grow old and feeble. If your battery has removable caps, you should periodically take them off and look inside with a good flashlight. Always carry out this chore with safety glasses on, and in a well-ventilated area. And don't smoke or bring any flame near the battery! The fluid inside should completely cover the metal plates. If it doesn't, you can add distilled water (not tap water) gradually until it does. Whenever you take your car into the shop, have the technician check the alternator to ensure that it's properly charging your battery when the engine runs.

Quick Question, Quick Answer

- How long will a typical automotive battery last before you need to replace it?
- Three to five years, usually, assuming that you take reasonably good care of it. But some batteries will fail inexplicably after a few months, while others will last for six years or more

Warning!　An automotive battery can release hydrogen gas, especially if it gets overcharged. This gas, along with oxygen, slowly bubbles out of the acid solution. As you learned in your high-school chemistry classes, hydrogen and oxygen combine violently to form water if they come near a spark or flame. Never smoke when working around an automotive battery. Never connect anything to the battery that could cause a spark to occur. Always work in a well-ventilated area so that any escaped gases can disperse.

Here's another reason why your car might not start. In order to get an internal combustion engine running, a battery must deliver a lot of current—far more than even the biggest appliance in your house demands. Because the voltage is low, it takes a huge current to muster the power necessary to crank the big engine in a car or truck. (Amps equal watts divided by volts.) For this reason, the wires leading to your battery must have large diameters, and the connectors must be clean at all times. In addition, the connections must have large metal-to-metal surface areas. A lot of electrons have to move through that metal, and a "bottleneck" anywhere, even at a single point, can keep your engine from getting the current surge that it needs to start up.

Did You Know?

Once in awhile, a battery will fail for no apparent reason. You might have some warning that it's getting weak, but sometimes it will die without warning. It'll start your vehicle normally when you leave home to go get some groceries, but when you come out of the store and turn the ignition key, nothing will happen at all.

Jump Starting

As long as your vehicle's alternator works properly, and as long as you drive your vehicle regularly, and as long as you avoid deep-freeze conditions, you can expect your automotive battery to maintain itself until it dies from old age. However, if you drive a lot, you'll eventually encounter a situation in which your vehicle won't start because the battery has lost its charge. In that case, you'll need to get a *jump start* from another vehicle. To do that, you'll either have to own a set of jumper cables (Fig. 5-2), or else find a helper who does. Once you've located somebody willing to give you a jump start, go through the following fifteen steps in order (Fig. 5-3).

Note that the instructions below apply only to *negative-ground* vehicles, which make up the vast majority of cars and trucks in the United States. (The term

FIGURE 5-2 A pair of jumper cables, roughly 6 feet (2 meters) long, with a gallon jug of washer fluid for size comparison. Note the large clips on the ends.

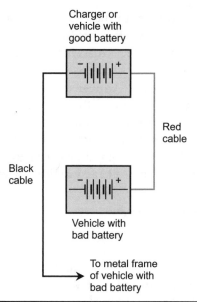

Charger or
vehicle with
good battery

Red
cable

Black
cable

Vehicle with
bad battery

To metal frame
of vehicle with
bad battery

FIGURE 5-3   Connection of bad automotive battery to a good battery for a "jump," or to a charger for replenishing. This diagram applies only to negative-ground vehicles (the vast majority in the United States).

"negative-ground" means that the negative battery terminal goes directly to the metal frame or chassis of the vehicle. The frame or chassis represents the reference voltage level, called "common" or "ground.")

1.   Open both vehicles' hoods and locate the batteries. Then close your helper's vehicle hood.
2.   Position your helper's vehicle so that the two batteries are as close together as possible. Then open your helper's hood again.
3.   If you have gloves, put them on.
4.   If you have safety glasses, put them on too.
5.   Unwind the jumper cables and straighten them out.
6.   Make sure that your helper's vehicle has its engine running.
7.   Attach the clip on one end of the red jumper cable to the positive terminal of the bad battery. Make sure it's secure and has a good electrical contact.
8.   Securely attach the clip on the other end of the red cable to the positive terminal of the good battery.
9.   Attach the clip on one end of the black cable to the negative terminal of the good battery.
10.   Connect the clip on the other end of the black cable to an exposed metallic object on the frame of the vehicle with the bad battery. Ideally, that object should be on the engine, but not close to the battery. Don't be surprised if a small spark occurs when you touch the clip to the frame.

11.   Turn the ignition key in the bad vehicle. It should start right away. If it doesn't, check to make sure that the electrical connections for the cables are secure. If you still have no luck, make sure that the clips on the ends of the cables are clean. The best ones will have a coppery color, like a new penny. If they don't, clean them with an emery cloth or steel wool.

12.   Remove the clip on the end of the black cable from the frame of your vehicle.

13.   Remove the other black clip from your helper's battery.

14.   Remove the red cable clips from both batteries.

15.   Keep your vehicle running until you can get it home (where, hopefully, you have a battery charger), or to a repair shop where they can replace the battery immediately!

Here's a Tip!

If you buy a pair of jumper cables now, put them in your trunk or in the back of your pickup's cab, and then keep them there indefinitely, you'll thank yourself some day. You can buy them for a few dollars at a good department store such as WalMart. Also get some safety glasses, a pair of gloves, and some emery cloth or steel wool. If you're really in the mood to spend money, you can buy a slow-charger for automotive batteries and keep it at home or in the vehicle.

Charging a Battery

In a motor vehicle, the alternator keeps the battery charged at or near full capacity all the time. You shouldn't have to use an "outboard" charger unless your battery has lost its charge for some reason. When you buy a new battery, it should be fully charged as you take it from the store, but you should top it off before you rely on it.

Figure 5-4   An "outboard" automotive battery charger with two different current settings. It's a little smaller than the battery itself (and a lot lighter).

To do that, you will need a charger such as the one shown in Fig. 5-4. The block diagram of Fig. 5-3 can serve as a guide for battery charging. The charger will have a red wire and a black wire, both with clips on the ends like the ones on the ends of jumper cables. Go through the following sixteen steps, in order, to charge your battery. Note that these instructions apply only to negative-ground vehicles.

1. Open your vehicle's hood.
2. Locate the charger so that the cables can easily reach the battery.
3. If you have gloves, put them on.
4. If you have safety glasses, put them on too.
5. Unwind the charger wires and straighten them out.
6. Plug the charger into a common utility outlet. Make sure that it's switched off.
7. Attach the clip on the end of the charger's red cable to the positive terminal of your battery. Make sure that it's secure and has a good electrical contact.
8. Attach the clip on the end of the charger's black cable to an exposed metallic object on the frame of your vehicle, reasonably far away from the battery.
9. Set the charger to the low-current mode and switch it on.
10. Leave the charger connected for several hours at the low-current setting.
11. Unclip the charger's black wire from your vehicle.
12. Unclip the charger's red wire from the battery.
13. Turn the ignition key. Your vehicle should start. If it doesn't, charge the battery for several more hours. If it still won't start, have your battery tested by an automotive expert.
14. If the battery tests show that it is bad, of course you should replace it.
15. Have your local auto expert check over your vehicle's entire ignition and electrical system right away, to make sure that the battery didn't go bad as a result of some other malfunction, or to correct any such problems if they do exist.
16. It's not a bad idea to replace a questionable battery even if it tests okay, especially if you can't blame its failure on any external cause such as a bad alternator.

Did You Know?

Chargers convert the AC from a utility outlet to a DC voltage that drives a controlled current through a battery's electrolyte solution. The best way to charge a battery is to use a relatively low current level, such as 2 A, and let the charger work overnight, or for however long it takes to ensure a full charge. If you charge a battery at 2 A, then you'll get 2 ampere-hours (2 Ah) after an hour's worth of charging, 4 Ah after two hours, 16 Ah after eight hours, and so on. In the extreme, if you have a brand new battery (say 80 Ah capacity) that hasn't been charged before, you'll need 40 hours at 2 A to charge it all the way up.

Your charger will probably have a setting that drives 10 A or more through the battery, as well as a low-current setting. In general, it's a bad idea to use the high-current setting in an attempt to charge a battery fast. Use the low-current setting, and have patience! I can back up this advice with direct experience. I once had a battery fail and got my truck home because somebody gave me a jump. I connected a charger to the battery at the 10 A setting, and it wouldn't work even after charging overnight. Then I switched the charger to the 2 A setting and, after six hours, I got the truck started. I immediately drove to the nearest WalMart, bought a new battery (fully charged right off the shelf), and installed it on the spot. When I got back home, I charged the new battery at the 2 A setting overnight to top it off.

For Nerds Only

Automotive batteries are meant to work at full charge, or nearly at full charge. However, you shouldn't take this principle to an extreme when you charge your vehicle's battery. Overcharging can cause *outgassing*, where hydrogen and oxygen bubble out of the electrolyte. If allowed to go on for a long time, overcharging can damage your battery. By the way, you should always charge batteries in well-ventilated areas so that if any outgassing does occur, the hydrogen and oxygen can disperse. You can place a small electric fan near the battery as you charge it, so any gases that do escape won't hang around the battery and create a fire or explosion hazard.

The Alternator

All modern cars, buses, and trucks have *alternators* that keep the battery charged and also help to run the electronic systems, lights, and optional devices, such as a sound system or two-way radio. In a typical car or light truck with an internal combustion engine, the alternator looks like a motor (Fig. 5-5) that's about the same size as a large grapefruit. You'll find it attached to the engine, or to some point under the hood near the engine. You'll see a drive belt attached to its shaft.

How It Works

An alternator is a small electric generator. Most alternators generate electricity by means of a complex magnet that rapidly spins inside of fixed wire coils; a few types work the opposite way, by rotating coils inside of fixed magnets. That's an oversimplification, but you should get the general idea. As the alternator turns, electric current arises in the wire coils as a result of a moving magnetic field. Engineers would tell you that *magnetic lines of flux* move with respect to the wire, so that the electrons in the wire are propelled to create the current. It's the same principle that makes the generators in a utility power plant work, but on a smaller scale.

Figure 5-5   The alternator looks something like a motor, and is usually located under the hood next to the engine.

The alternator in a car or light truck can generate several amperes of current when the engine runs at normal driving speed. The current, and also the frequency of the AC that the alternator generates, vary depending on the engine speed. A rectifier circuit, comprising a set of semiconductor diodes, changes the AC to DC for operation of the vehicle's electronics. A special voltage regulator keeps the output voltage of the alternator/rectifier combination at a level slightly greater than the battery voltage, typically 13 V or 14 V.

When your vehicle's engine runs, the alternator charges the battery so that the battery doesn't deplete when you operate onboard devices, such as a radio, air conditioner, heating fan, headlights, or two-way transceiver. The vehicle computer consumes electricity, too. The alternator gets its spin from a drive belt connected to the engine. If the drive belt breaks or slips, or if something goes wrong inside the alternator, or if an electrical connection associated with the alternator happens to break, the battery will bear the entire burden of running those onboard electronic devices, and pretty soon you'll find yourself looking for a jump!

When It Fails

An alternator can go bad suddenly and totally (engineers call that sort of event a *catastrophic failure*). However, an alternator can fail gradually as well, getting less and less effective until you find out about it one way or another.

If your vehicle has a voltmeter on the dashboard, you should check that meter's reading often while the engine runs. It should normally indicate at least 12 V; sometimes (especially right away after you start the engine) it will rise to as much as 14 V. If it falls below 12 V or rises much above 14 V, you should suspect that something's wrong with the alternator. The notable exception is the short interval between the moment you first turn the ignition key and the instant the engine starts. You should expect the voltmeter reading to drop considerably below 12 V until the engine starts up.

An alternator can fail internally, giving you no visible sign except the voltmeter reading on the dashboard. You can check that voltage with a multimeter such as the type you learned about back in Chap. 2. A digital meter will work better than an analog meter for this test. Set the meter for a DC voltage range that will make it easy for you to see readings from about 10 V to 15 V. (In the analog meter at left in Fig. 2-10, that would be the 0–50 V DC switch position with the rotary switch pointing "northeast"; in the digital meter at right, it'd be the 0–20 V DC position with the switch pointing "northwest.") Then go through the following procedure:

1. Put on a pair of safety glasses.
2. Put on a pair of gloves.
3. Double-check to make sure that the meter is set to measure DC voltage (not current or resistance!).
4. Make sure the engine is off.
5. Make sure all of the vehicle's lights and accessories are off.
6. Open the hood.
7. Check the battery voltage by holding the meter's black probe tip against the negative battery terminal and the meter's red probe tip against the positive battery terminal.
8. You should see a reading of 12 to 13 V.
9. Start up the engine.
10. Check the battery voltage again. It should be slightly greater than it was with the engine off.
11. Turn on all the vehicle's accessories and lights, and keep the engine running.
12. Check the battery voltage a third time. You should not see any significant drop in the battery voltage compared with the second reading.
13. If you see a significant voltage drop with the lights and accessories on and the engine running, take your vehicle to a shop to have the alternator tested and replaced if necessary.

Sometimes the drive belt, which connects the engine mechanically to the alternator, will slip. You might not be able to see or hear any evidence of it, but it will reduce the effectiveness of the alternator, especially under heavy load (with all the vehicle's accessories and lights on). Sometimes you'll hear a soft squealing or hissing sound coming from the point where the drive belt connects to the alternator. Of course, if the drive belt has broken and is, therefore, missing, you'll notice right away! (In Fig. 5-5, the belt is clearly visible. If it broke, the visual change would be hard to miss.) The solution in a case of this sort is simple: Have that belt replaced by a competent auto mechanic. Then have the alternator itself tested and replaced if necessary.

Here's a Tip!

Unless you happen to be an auto mechanic, don't try to replace a bad alternator yourself. Take your vehicle to a repair shop that you trust, and have them do it. They'll know exactly what sort of alternator to install, how to ensure that it's wired up right, and how to test it after installation.

Lights

Have you ever counted all the lights and lamps in your vehicle, including the ones on your dashboard? You ought to do it someday. You'll be amazed! In a modern car or truck, most of the lamps are installed in such a way that you can't replace them yourself unless you're an auto mechanic. Motor vehicles employ two types of lamps: *incandescent lamps* and *light-emitting diodes* (LEDs). In older vehicles, most of the lights comprise incandescent lamps. Over the years, LED technology has improved, so LEDs have gradually replaced incandescent bulbs in most applications.

Incandescent Lamps

An incandescent lamp works by allowing an electric current to flow through a piece of wire that has a precisely tailored resistance and current-carrying capacity. As a result, the wire, called the *filament*, glows white hot. The bulb has a transparent, airtight case called the *envelope*. The typical incandescent lamp has a filament made of fine tungsten wire, which is often wound into a "coiled coil" configuration. The manufacturer spins the wire into a tight, long *helix* (coil), which is then wound into a larger helix. This geometry maximizes the length of wire that can fit inside the envelope, ensuring the greatest possible amount of light emission. The interior of the envelope contains either a complete vacuum or a rarefied (low-pressure) inert gas. Figure 5-6 shows the basic construction of a small incandescent lamp, such as the sort that you'll find in an older automobile's turn signals and headlamps.

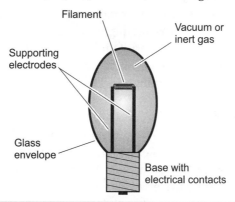

FIGURE 5-6   Anatomy of a small incandescent lamp for automotive use. It measures approximately 3/8 to 3/4 inch (1 or 2 centimeters) tall.

Did You Know?
You can see the "coiled coil" in any incandescent bulb that has a clear envelope and that takes advantage of this filament geometry. Examine the wire through a magnifying glass when the filament is not glowing (when the bulb is off).

The current that flows through an incandescent bulb's filament depends on the voltage that you put across it, and also on the end-to-end resistance of the filament wire. The filament resistance, in turn, depends on the temperature. A cold filament has lower resistance than a hot one. When you first apply voltage across the filament, a lot of current flows through it, so it warms up fast. As its temperature rises, the resistance increases, reducing the current, so the filament reaches a stable maximum temperature at which it glows white hot.

Incandescent lamps are notorious for their low efficiency. The typical incandescent lamp converts only about 10 percent of the applied power to visible-light output. The remaining 90 percent of the input power radiates from the device in the form of *infrared* (IR) rays, often called "heat." For this reason, energy-efficiency advocates have condemned incandescent lamps for many years, and now that more efficient lamps are available, automakers are gradually converting over to them and doing away with incandescent bulbs.

Traditional incandescent lamps suffer from short life spans as well as relative inefficiency. Tungsten slowly evaporates from the filament, limiting its useful life. The vaporized tungsten condenses on the inside of the glass envelope, gradually darkening it and further reducing the lamp's efficiency. These problems can be overcome to a large extent by filling the envelope with *halogen vapor* at low temperature. Iodine is the most common halogen used for this purpose, although chlorine and bromine have also been employed.

Figure 5-7 shows the internal anatomy of a small halogen lamp. The envelope is made of quartz rather than glass. A halogen lamp filament carries more current, and therefore, operates at a higher temperature than a conventional incandescent lamp does. Quartz can withstand higher temperatures than glass can tolerate. The higher filament temperature results in improved efficiency and "whiter" light at maximum current, compared with standard incandescent lamps. Halogen lamps work well as automotive headlights.

Semiconductor LEDs

The term *semiconductor* arises from the ability of certain materials to conduct some of the time, but not all the time. Whether or not they conduct depends on the way that the materials are treated during the manufacturing process, and also on the voltages that you apply to electrodes embedded in the materials. Various elements, compounds, and mixtures can function as semiconductors including *silicon, gallium*

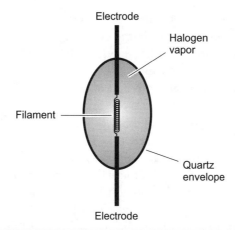

Electrode

Halogen vapor

Filament

Quartz envelope

Electrode

FIGURE 5-7 Anatomy of a halogen lamp, such as might be used in an automotive headlight. It measures approximately 3/4 inch (2 centimeters) tall.

arsenide (abbreviated GaAs), *germanium, selenium, cadmium* compounds, *indium* compounds, and the oxides of various metals.

Impurities, also called *dopants,* give a semiconductor material the properties that it needs to function as an electronic component, and therefore, as a light-emitting device. The impurities cause the substance to conduct current under specific conditions. When manufacturers add an impurity to a semiconductor element, they call the process *doping.*

When a dopant contains an excess of electrons, we call it a *donor impurity.* A semiconductor with a donor impurity is known as *N type material* because an electron carries a negative (N) electric charge. If an impurity has an inherent deficiency of electrons, we call it an *acceptor impurity,* and the resulting material conducts mainly by means of *hole flow.* A *hole* is a location within an atom where an electron "should exist, but doesn't." A semiconductor with an acceptor impurity is called *P type material* because a hole has a positive (P) electric charge.

In a semiconductor component called a *diode,* a wafer of P type material is placed into direct contact with a wafer of N type material, and electrodes are embedded in each wafer for connection to external circuits or voltage sources. When the N type material has a sufficient negative voltage with respect to the P type, electrons flow easily across the barrier (called the *P-N junction*) from the N type wafer to the P type wafer. The N type substance constantly "feeds" electrons to the P type substance in an "attempt" to create an electron balance, and the voltage source keeps "feeding holes" to the P type substance in order to sustain the electron imbalance. This condition is called *forward bias.*

An LED is specially designed to emit visible light when it is forward biased so that current passes through it. The current energizes the atoms in the vicinity of the P-N junction, causing the electrons to temporarily rise to unnaturally high orbits around the nuclei. The electrons inevitably fall back from these artificially high orbits into normal orbits. When the electrons fall back, they lose the energy that they previously gained by radiating visible light, so the junction glows.

One of the most significant aspects of LED technology is the fact that these devices produce a reasonable amount of light but consume only a small amount of electrical power. As a result, LEDs cost less to operate, once you buy them, than any other type of lamp known (as of this writing). By their nature, LEDs are low-voltage devices, and they require DC to work. They work best with about 3 V of forward bias. If you connect four LEDs in series across a 12.6-V automotive battery, then each LED will receive a little bit more than 3 volts, which will do quite well.

The only trouble with the foregoing voltage-regulating arrangement lies in the fact that if a single LED in a series chain burns out, the whole set will go dark. Engineers get around this bugaboo by connecting two or more LEDs in parallel (directly across each other, like the rungs in a ladder) for each spot in the series connection. That way, if a single LED fails, its parallel companion(s) can keep current flowing through the main series chain. Only the bad LED will go dark, instead of the entire congregation going out.

For Nerds Only

A typical LED bulb will work for about 30,000 hours if you operate it under reasonable conditions. According to that specification, if you use it for an average of six hours a day, you can expect it to last for over 13 years. Of course, some LED devices will burn out sooner than that, and some will last longer.

Fuses

All vehicles have fuses that protect the electrical circuits and devices from overload. The fuses also protect the battery and the alternator in case of a short circuit. Fuses minimize the risk of electrical overheating that can cause fires (and in a vehicle, a fire is the last thing you want). Your headlights, brake lights, turn signals, backup lights, interior lights, climate-control fan, radio, and other electrical devices all have fuses in their lines. If a fuse blows, the affected device or circuit won't get the electricity it needs, so it can't work.

What They Look Like

Most newer cars have *blade fuses*. Older vehicles sometimes have *cylindrical fuses*. Figure 5-8 shows two blade fuses and two cylindrical fuses along with a U.S. quarter dollar for size comparison. All fuses have an amperage rating, which tells you how much current they can carry before they blow out. They also have a maximum voltage rating, such as 24 V or 48 V.

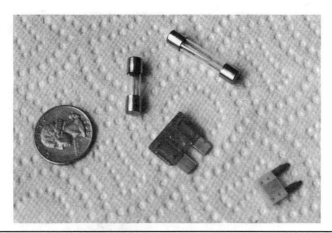

FIGURE 5-8   Four typical automotive fuses next to a U.S. quarter-dollar coin for size comparison.

A good fuse, either the blade or cylindrical type, has a continuous, solid wire visible inside. If the wire looks intact, the fuse is good. If it looks broken or burned up, the fuse is bad. If you can't tell for certain whether or not that little wire in a fuse is broken by simply looking at it (some fuses are cloudy or opaque), you can test the fuse with a multimeter set to measure ohms. A good fuse will show zero ohms (a short circuit) between its two terminals. A bad fuse will show "infinity" ohms (an open circuit).

The Fuse Panel

In order to locate the fuse panel in your vehicle, you might have to consult the owner's manual. Alternatively, you can search around under the dash and inside the door jambs with the doors open. A few vehicles have the fuse panel underneath the hood near the engine. If you've lost your vehicle's instruction manual and can't find the fuse panel, you can have your friendly local auto technician show you where it is.

Figure 5-9 shows the fuse panel in my vehicle, a 2003 Chevy S10 pickup (at the time of this writing), also known as "Old No. 7." The panel resides in the driver's-side door jamb. I have to take a protective cover off the panel to access the fuses. The slots are numbered from 1 to 24. In order to know which slot goes to which circuits or devices, I must consult the owner's manual. The numbers on the fuses themselves, as they appear in this photograph, are the amperage ratings. My fuse box has five empty slots, which can be used for connections to auxiliary electronic devices, such as hi-fi sound systems or two-way radios.

How To Replace a Blown Fuse

You can remove a blade fuse using small needle-nosed pliers. Just pull the whole component straight out. You can pull out a cylindrical fuse with your fingers (usually) but you should wear gloves in case the glass breaks. To replace either type

FIGURE 5-9   Fuse box in a typical motor vehicle with the protective panel removed. In this case (a 2003 Chevy S-10 pickup), the fuses are accessible only with the driver's side door open.

of fuse, you can simply press the new component back into place. Of course, the new fuse will have to be exactly the same physical size as the old one! And again, it should have the same amperage rating as the old one.

Quick Question, Quick Answer

- What will happen if I replace a blown-out fuse with one that has a different amperage rating?
- If the new fuse has a higher amperage rating than the old one, it won't adequately protect the circuit in which it's connected, and you'll run the risk of overheating and possibly even having a vehicle fire. If the new fuse has a lower amperage rating than the old one, it will probably blow either right away or after a short time, causing you a lot of unnecessary inconvenience.

Two-Way Radios

Since the end of the twentieth century, cell phones have almost eliminated two-way radios in consumer vehicles, with the exception of amateur (ham) radio operators who still use them for hobby and for emergency communications. Nevertheless, in some parts of the United States and much of Canada, cell phones won't work because there isn't any repeater tower within range. If you regularly drive, or plan to drive, through a zone where cell phone coverage isn't available, you might want to think about installing a two-way *mobile radio transceiver* in your vehicle.

Types of Radios

Are you old enough to remember the *Citizens Band* (CB) radio craze of the late 1970s and early 1980s? This mode became popular among truckers and motorists who refused to obey the national 55-mile-an-hour speed limit imposed in the wake of oil market shocks. People employed their "CBs" to warn each other about speed traps. When the 55-mile-an-hour speed limit was repealed in the 1980s, CB radio use declined.

Other factors besides the repeal of the unpopular speed limit had an effect on CB use through the late 1980s and into the early 1990s. The *skip* phenomenon, which occurs at 11-year intervals during sunspot-cycle peaks and can cause CB signals to travel thousands of miles, frequently rendered the radios useless for local daytime communications. You might find a signal from Peru overriding a signal from a few miles down the road, and all 40 CB channels would fill up with a cacophony of conversations. The explosion of cell phone popularity in the 1990s delivered the final blow. Today in the United States, CB usage is confined to businesses, ranchers, and a few folks who enjoy its unique cultural flavor. The mode remains popular in some parts of Central and South America.

Amateur radios resemble CB transceivers, but operate over a wider range of frequencies with more powerful transmitters, offering better performance. For the several hundred thousand people in the United States who have government-issued amateur radio licenses, "ham radio" is a clear winner over CB radio for usability and versatility. The Morse code requirement, which in the olden days kept a lot of people out of the hobby, no longer exists. You must take a test to get an amateur radio license, but the entry-level license exam is easy for anybody with a little electronics knowledge to pass. For information on how to get an amateur radio license, visit the website of the American Radio Relay League at www.arrl.org.

If you want to buy a CB radio, I recommend Radio Shack stores, which (as of this writing) carry CB transceivers and antennas to fit any budget. You don't need a license to operate one of these radios, although strict legal limitations exist on power, antenna systems, and general usage. The instruction manuals for most CB radios will tell you about these matters. Otherwise, you can enter "CB radio" or "Citizens Band" as a phrase in your favorite Internet search engine and read the contents of the top hits. You can also read about CB radio in Wikipedia.

Whatever type of two-way radio that you decide to buy (if any), you should mount it securely in your vehicle, preferably under the dashboard. Figure 5-10 shows the author's mobile communications system, which comprises an amateur radio transceiver with a CB transceiver directly underneath it. The "ham radio" works at 144 to 148 megahertz (MHz), also known as 2 *meters* because that's the electromagnetic wavelength of the radio waves as they travel though the atmosphere.

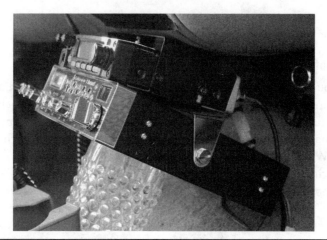

FIGURE 5-10   Two-way radios mounted under the dashboard. The amateur radio is on top and the CB radio is on the bottom. A cup, wedged underneath the CB radio, keeps both units from wobbling.

The CB radio works at 27 MHz, also called *11 meters*. The "ham radio" is mounted to the dashboard with a bracket supplied by the manufacturer. The CB radio and its mounting bracket are attached to the bottom of the ham radio with Velcro strips. The whole assembly is held in place by a plastic cup wedged underneath the CB unit. I never use both units at the same time, of course, so mutual overheating never presents a problem.

Did You Know?

Specialized business radios are available if you need them. They work like CB and amateur radio sets, and the same installation rules and precautions apply.

Power Connections

Most two-way mobile radios are designed to work with a power supply that delivers 12 to 14 V DC, which coincides with the voltage in a standard motor vehicle in the United States. A two-way radio will normally come supplied with a power cord that has a red wire for the positive connection and a black wire for the negative connection. The power cord will also have one or two fuses in it, usually of the cylindrical type and installed in plastic holders.

The ideal way to connect a two-way radio to you motor vehicle's electrical system involves running the radio's power wires directly to the battery terminals. If you can't do that, or if you don't want to drill any holes in the firewall that separates the vehicle interior from the engine compartment, you can connect the wires into a vacant pair of slots in the fuse panel. You must keep the fuse or fuses intact in the radio's power cord, however! A good automotive supply store can provide you with a special adapter that will let you connect the power wires to the fuse panel.

A third method of power connection involves the use of an adapter that you can insert into the vehicle's lighter socket, which you'll usually find on the dashboard. Figure 5-11 shows the author's power cords, cut to length (but again, making sure that the fuses stay there) and attached to lighter adapters available at Radio Shack stores or on the website at www.radioshack.com. I used the connectors with banana jacks and screw terminals made by Enercell; as of this writing that item carries the Radio Shack part number 270-036.

Antennas

No radio, whether fixed or mobile, can work well without an efficient antenna designed for use at the intended frequency. For CB radios, you can buy a variety of antennas ready-made from Radio Shack stores, and also from some hardware stores and trucking supply outlets. The typical CB antenna has a helix of wire wound on a fiberglass rod measuring anywhere from 2 to 6 feet long. With the shorter antennas, you can use a magnetic mount ("mag mount") if your vehicle has a steel sheet-metal frame. The best place to put the antenna is in the center of the rooftop, as shown in Fig. 5-12. The second best place, for conventional cars, is on the top of the trunk. Figure 5-13 shows a close-up view of my magnetic mount on the roof of my Chevy S-10.

If you want to use an antenna longer than 3 feet, you should rely on a mounting scheme that provides a secure mechanical bond to the vehicle. A large antenna with a magnetic mount will blow off at high speeds because of the wind. (Think about

Figure 5-11 Power plugs for the radios, with lighter adapters on the ends. The power cords contain separate fuses to protect the radios and the vehicle's electrical system.

FIGURE 5-12   A CB mobile radio antenna attached to the roof of a pickup truck with a magnetic mount.

it. On a freeway where the speed limit is 65 miles an hour, if you drive at that speed into a 10-mile-an-hour headwind, you'll subject your antenna to the equivalent of a category 1 hurricane.) Various antenna mounting brackets exist for large mobile antennas; once again, Radio Shack stores are a great place to find them. The most effective CB radio antenna is a stainless steel whip that measures 102 inches (259 centimeters) long. It doesn't involve any artificial shortening, such as helically wound antennas do, so it will produce the best reception and transmission efficiency that you can expect from any CB mobile radio antenna.

FIGURE 5-13   Close-up view of a magnetic mount suitable for small antennas. It needs a ferromagnetic surface (such as steel sheet metal) in order to stay in place.

If you have an amateur radio set, the same antenna rules apply as with CB antennas, insofar as the type of mount that you should use. A radio designed for operation at 144 MHz or above will have an antenna short enough for use with a magnetic mount. But if you plan to install a radio for use at any amateur radio band "below 2 meters," that is, at any frequency below 144 MHz, you'll need a specialized antenna along with a heavy-duty mounting bracket designed especially for it. You can find lots of advertisements for such antennas in a magazine called *QST*, published especially for amateur radio operators. Most large towns have amateur radio supply stores. You can usually find suitable mobile antenna hardware at those places. They might ask you to present your amateur radio license to make a purchase, however, so take along a copy of that license!

Noise and Interference

The most frequently encountered problem with mobile transceiver installations, especially CB transceivers and "ham radios" on the lower frequencies (below 30 MHz), is interference to your receiver from the vehicle's engine and charging systems. *Ignition noise* can completely drown out all the signals on some frequencies, and *alternator whine* can prove an annoyance as well. Ignition noise sounds like an endless series of rapid-fire "pops" or a steady "buzz," while alternator whine creates an audible tone whose pitch goes up and down as you accelerate and decelerate. The *noise blanker* in your radio's receiver can reduce or eliminate ignition noise in some cases, but it won't do any good for alternator whine. If you have bad reception with a mobile radio installation for either of these reasons, take the vehicle to a good shop. They might be able to modify your vehicle to reduce the problem without compromising engine performance.

Another potential problem involves the transmitter interfering with your vehicle's computer. If your transmitter produces more than a few watts of radio-frequency (RF) output, its signal can make your computer do some strange things. I recall a situation where transmitting a signal would instantly cause my engine to stop. Obviously, that state of affairs presented a danger! Computer-related RF problems can sometimes be eliminated by running the power-supply wires from your radio directly to the vehicle battery, and also installing little donut-shaped powdered-iron devices called *Ferrite beads* at several points along the radio's power wires. Another way to prevent this type of interference is to reduce your transmitter's output power. Yet another solution is simple and always works: Don't transmit while your vehicle moves!

Warning! In some states and towns, you can't legally use electronic equipment such as cell phones, mobile radios, and computers while driving a vehicle. You should check with your local law enforcement people to make sure that you'll stay within the law if you use a mobile radio transceiver while driving. If you're traveling, you had better make sure that you don't arouse the unwanted interest of overzealous cops or state troopers by using mobile equipment where it's illegal!

Hi-Fi Sound Systems

Do you remember when the only sound-producing device in a car or truck (other than restless kids or pets, maybe) was a monaural AM radio? Now you can listen to dozens of satellite radio stations or play music from little chips that hold thousands of tunes, and pump high-quality digital sound into speakers that give you the impression of sitting in a concert hall.

Think and Plan

Assuming that you don't already have a sound system that satisfies your desires, you had better think for awhile before you spend hundreds of dollars on mobile audio equipment. Here are some factors to consider, in the form of questions that you can ask yourself and then honestly answer.

- How old is your vehicle? How much longer do you plan to keep it?
- Do you plan to leave the sound system in the vehicle when you sell it? Or will you take the sound system out and put it in your next vehicle?
- Do you want to sell your vehicle with holes marking the places where the components of an old sound system were?
- How much ambient noise does the vehicle produce when you drive at freeway speeds? Do you get a quiet ride, or does it sound like a storm?
- How loud do you want the sound from your system to be?
- What sorts of music do you, and your significant others, prefer? Classical and soft rock don't need as much "oomph" (bass) as hip-hop or heavy metal does.

Once you've decided what you want, you would do well to consult a reference book devoted entirely to mobile audio systems so you can have an intelligent conversation with a salesperson. Alternatively, you can take a knowledgeable audiophile friend with you to serve as your interpreter. (I assume that you aren't reading this section if you're a mobile audio expert already.) Don't let a fast-talking salesperson "snow" you. If you get bad impressions from the people at a particular shop, take your business somewhere else. Does this seem like gratuitous advice? Well, ask yourself: How many times have you let a slick salesperson talk you into buying something other than what you came into the store to get in the first place?

For Nerds Only

Serious mobile audiophiles can get plenty of information in Andrew Yoder's book, *Auto Audio* (McGraw-Hill, 2000). Digital audio recording and reproduction technology has advanced since that book was published, but many of the technical and troubleshooting aspects of mobile hi-fi systems, especially concerning amplifiers, speakers, and wiring, remain valid today.

The Head Unit

Figure 5-14 is a block diagram of a complete automotive sound system. Let's start with the items inside the gray-shaded box. The combination of all these devices forms the *head unit*. It gathers signals from the radio, and/or converts data from media, such as compact-disc (CD) players, MP3 players, or tape cassettes, into audio signals. When you buy a vehicle, it will have a factory-installed head unit that makes up the heart of the so-called *stock system*. My venerable "Old No. 7" pickup has a head unit that includes an AM/FM radio and a compact-disc (CD) player. That's all it has and that's all I need because I don't drive very much (and would just as soon never drive at all). A factory-installed head unit sits in the dashboard, near the center, so that the driver and the front-seat passenger can both conveniently reach it.

If you want to install a custom sound system in your vehicle, you'll most likely want to start with the basics and work your way up, so you'll have to shop for a new head unit. What types of media do you plan to use? Old-fashioned tape cassettes, or CDs, or MP3 media, or satellite radio, or some combination of them, or all of them? A good sound-system dealer that specializes in "car stereo" systems can help you out here.

Do you want your system to improve on the sound that you get on your stock unit FM radio, so that you'll quit using the stock unit altogether? If so, your new head unit should have a built-in FM receiver. Does AM radio matter to you? Maybe

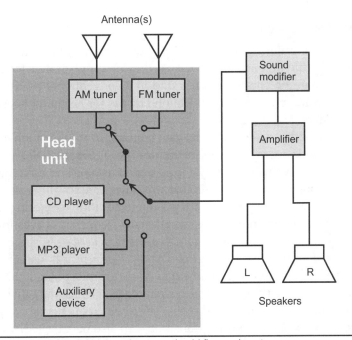

FIGURE 5-14   Components of a high-end automotive hi-fi sound system.

you can remove the stock unit and replace it with a new head unit, although you'll probably have to modify the hole in the dashboard to accommodate the new head unit. Whatever you decide, don't scrimp on the head unit. The quality of the signal that comes out of an audio system's amplifier can't be any better than the quality of the signal that goes into it. And a head unit that's hard to "work" while driving can actually put your life at risk.

Here's a Tip!

Once you find a head unit that meets your needs, see if you can get the dealer to let you see it in action. Then ask yourself the following questions, and make sure that you're satisfied with the answers that your eyes provide.

- Can you easily identify and manipulate the controls?
- Can you read the display in bright, direct sunlight?
- Can you read the display in total darkness?
- Can you read the display when your eyes are above it or off to either side?
- Are the display characters large enough so that you can read them from three or four feet away?
- Do you like the display colors?
- Do the individual knobs, buttons, and switches have lights that make them easy to identify in total darkness?

Sound Modification

Sound modification devices usually have circuits that change the frequency response of the system. In lay terms, they're sophisticated tone controls, more elaborate than the ones that you'll find in a conventional head unit or amplifier. The most common sound modification circuits for automotive use are *shelf filters* and *equalizers*.

In a *low-frequency shelf filter*, the gain (amount of amplification) is *unity*, meaning that the device doesn't amplify or reduce the signal level above a certain critical frequency. Below that frequency, the gain is adjustable and can range over negative as well as positive values. Therefore, the filter can act either as a low-frequency amplifier or as a low-frequency attenuator, in addition to offering a flat response, if desired. Figure 5-15 shows some gain-versus-frequency curves for a hypothetical low-frequency shelf filter, with gain expressed in *decibels* (a special way to express sound volume or signal strength). The *shelf frequency* is usually defined as the frequency at the center of the curve transition (the mid-point of the sloped portion). However, it can also be defined as the frequency at which the gain begins to increase or decrease from its unfiltered level. The shelf frequency is adjustable. In some filters, the *skirt slope*, or rate of transition, can also be adjusted. You can think of a low-frequency shelf filter as a sophisticated form of bass tone control.

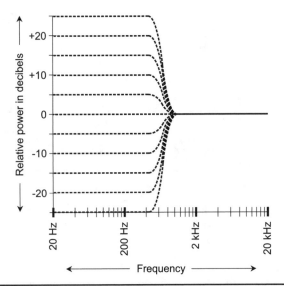

FIGURE **5-15**   Some gain-versus-frequency curves for a hypothetical low-frequency shelf filter. Dashed curves denote the shelving effect.

For Nerds Only

When engineers calculate or determine the loudness of a sound in decibels, they take into account the fact that our ears and brains perceive sound volume according to the logarithm of the actual sound power, not directly with the actual sound power.

In a *high-frequency shelf filter*, the gain is unity below a certain critical frequency. Above that frequency, the gain can be continuously varied. The shelf frequency is also adjustable. The skirt slope can be varied in some devices. This type of filter can act either as a high-frequency amplifier or as a high-frequency attenuator, in addition to producing a flat response. Figure 5-16 shows a family of gain-versus-frequency curves for a hypothetical high-frequency shelf filter. You might imagine it as a sophisticated treble tone control. A low-frequency shelf filter and a high-frequency shelf filter, when combined into a single device, can act as a versatile tone control.

A *graphic equalizer* is a device that lets you adjust the relative loudness of audio signals at various frequencies. It allows for meticulous tailoring of the amplitude-versus-frequency output (in other words, the tone) of hi-fi sound equipment. The circuit contains several independent gain controls, each one affecting a different part of the audible spectrum. The controls usually take the form of *slide potentiometers* with calibrated scales. The slides move up and down or left to right. When you set the potentiometers so that the slides are all at the same level, the audio output or response is flat, meaning that no particular range is amplified or attenuated (quieted down) with respect to the other ranges. By moving any one of the controls, you can

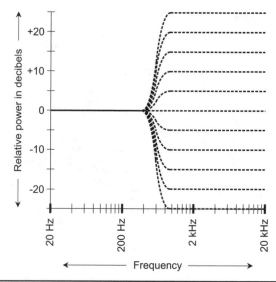

FIGURE 5-16 Some gain-versus-frequency curves for a hypothetical high-frequency shelf filter. Dashed curves denote the shelving effect.

adjust the gain within a certain frequency range without affecting the gain outside that range. The positions of the controls on the front panel provide an intuitive graph of the output or response curve. Figure 5-17 is a block diagram of a hypothetical graphic equalizer with six gain controls.

Figure 5-18 is an example of a gain-versus-frequency characteristic (gray solid line) that can be obtained using the equalizer circuit diagrammed in Fig. 5-17. The individual bandpass responses are shown as dashed curves. This particular equalizer has six filters to keep the illustrations simple, but in high-end professional audio

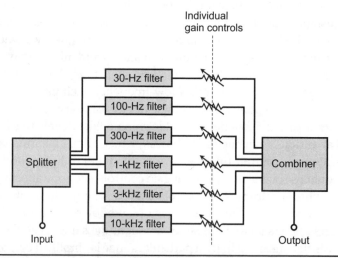

FIGURE 5-17 Block diagram of a graphic equalizer with six bandpass filters.

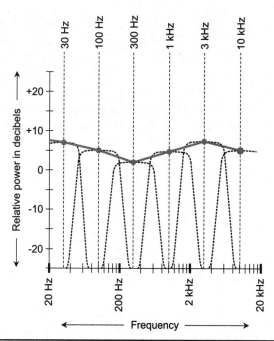

FIGURE 5-18 An example of a gain-versus-frequency characteristic (gray solid line) that can be obtained using the circuit shown in Fig. 5-17. The individual bandpass responses appear as dashed curves.

work, you'll sometimes come across equalizers with 20 or more individual filters for each audio channel in the system. Car stereo units might have as many as 10 or 12 filters.

A graphic equalizer in which the gain, center frequency, bandwidth, and skirt slopes are independently adjustable for each filter is known as a *parametric equalizer*. In addition to frequency-specific filters for each channel, a parametric equalizer can incorporate a low-frequency shelf filter and a high-frequency shelf filter, both of which have adjustable gain, shelf frequency, and skirt slope. A well-designed parametric equalizer can produce any gain-versus-frequency characteristic an audiophile could possibly want, and can compensate for the acoustic characteristics of practically any reproduction or performance venue, from a concert hall to an old truck like mine!

The Amplifier

If you're serious about having a good sound system in your vehicle, you'll need a dedicated amplifier. Some head units have amplifiers, but in most cases they won't provide enough *peak power* to guarantee excellent hi-fi sound quality throughout the entire range of frequencies that humans with "good ears" can hear (20 Hz through 20 kHz), or better yet, over an extended range of frequencies (about 10 Hz through 40 kHz).

The peak power output of an amplifier is, as the term implies, the amount of audio power that you get at sound peaks. The peaks occur only for extremely short periods of time, so you'll sometimes hear engineers talk about *instantaneous power* levels. The peak power at any particular audio frequency is the maximum instantaneous power at that frequency. Bass peaks typically occur at different times than midrange peaks, which in turn occur at different times (or instants) than treble peaks. For most applications, an amplifier that provides 100 W peak output will work well enough.

Did You Know?

If you want good sound from any hi-fi system, whether it's in your vehicle or in your home, you'll need an amplifier that will provide distortion-free performance at peak power levels considerably greater than the actual peak power levels you expect to have in everyday use. In other words, you'll want to take advantage of the well-known tactic known as *overengineering*.

The amplifier follows the sound modifier (if any) in an audio system, as shown in Fig. 5-14. The amplifier has two channels, as do the sound modification system and the head unit. These channels are normally called "left" (L) and "right" (R). Both channels should provide the peak power level that you desire. The average power level will be considerably less than the peak level, but in any case, a good amplifier will place a considerable load on the vehicle's electrical system.

Any good audio amplifier will produce quite a lot of heat because all hi-fi systems operate at an *efficiency* of only 40 to 50 percent. Strangely enough, improved sound quality comes at the expense of efficiency, so the best-sounding hi-fi systems are often the least efficient ones! If an amplifier can produce 100 W of peak audio output power at an average of 20 W per channel, then you'll end up with 50 to 70 W (roughly) of power input per channel, so your system will produce between 30 (50 W in minus 20 W out) and 50 (70 W in minus 20 W out) of heat per channel. You could easily get 100 W of heat coming off of the amplifier unit.

For Nerds Only

You can calculate the efficiency of an amplifier by dividing the average audio output power by the total average input power, expressing both figures in watts. Sometimes you'll see the term "RMS" (for root-mean-square) instead of the term "average," even though technically, the two terms don't mean exactly the same thing. The average input power (in watts) is the electrical system's voltage (in volts) times the amplifier's average current demand (in amperes). The efficiency will always turn out as a ratio of less than 1. To get the efficiency as a percentage, multiply the efficiency ratio by 100.

All substantial audio amplifiers have *heatsinks* comprising solid, massive, finned metal structures that are attached to the amplifiers, usually on the back panel or on the bottom. You must make sure to locate your amplifier so that the heatsink gets good air circulation, and isn't near anything combustible. The trunk is a good place. The engine compartment is another. It's a bad idea to put an amplifier beneath a seat, however, because the air circulation won't provide enough cooling, and we all know how combustible stuff can accumulate down there.

In order to ensure the least possible distortion in your amplifier, you must make sure that it gets a constant voltage as the current demand varies. You should have a technician check over your vehicle's electrical system to ensure that it can handle the load that the amplifier will place on it during sound peaks. If the voltage at the amplifier terminals drops during sound peaks, they'll be "blunted" as they reach the speakers, and you won't get the full sound quality that the amplifier can theoretically provide. Also, you should use the thickest possible wire to connect the amplifier to the electrical system, and make certain that all the individual connections are clean, tight, and substantial.

The Speakers

Of all the components in a sound system, the speakers make the most "per dollar" difference in the ultimate sound quality. If you have good speakers, you'll probably have good sound; if you have bad speakers, you'll *inevitably* have bad sound. Speakers are passive devices, meaning that they don't need any power supply in order to function. They convert the amplifier's output into sound waves without the need for a battery or source of utility power.

Speakers for automotive hi-fi systems are generally smaller than their counterparts in home systems, simply because there's a limited amount of physical space available in which to mount them. This size constraint doesn't affect the highest audio frequencies (treble), and only affects the midrange frequencies to some extent. However, the size constraint makes it difficult to get truly good low-frequency (bass) reproduction. The stock system's speakers in a typical car or truck each comprise a single physical assembly for the entire range of audio frequencies.

In some regions of the United States, the interior temperature of a vehicle's enclosed cabin can fall below −10°F (−23°C) on bitter northern winter nights and rise to more than 140°F (60°C) on sunny summer afternoons. Such wild temperature variations, along with changes in relative humidity, can ruin the cones in ordinary speakers in less than a year. The best automotive speakers have specially treated cones to minimize the impact of the environment on long-term performance. Nevertheless, if you spend a lot of money to put a top-notch sound system in your vehicle, you should do everything that you can to keep the cabin temperature from reaching extremes, even for short periods of time.

The placement of your speakers matters as much as the quality of the speakers themselves. You should try several different arrangements to see which one gives the best results in your particular vehicle. Once you have an arrangement that satisfies you, then you should anchor your speakers so that they won't move (or fly around!) inside the vehicle.

Did You Know?

Your vehicle's cabin acts like an echo chamber at some audio frequencies and like a wide-open space at other frequencies. This phenomenon, called *resonance effect*, can produce unexpected and bizarre results. For example, a system that sounds fabulous to a short driver who has to put the seat all the way forward might sound terrible to a tall driver who has to put the seat all the way back.

Electrical Noise

Electrical noise is produced by discrete components in electronic equipment, such as transistors, diodes, resistors, and integrated circuits (ICs). Minimizing the generation of this noise is a major priority in the design of hi-fi audio systems, particularly in circuits or devices that are followed by multiple stages of amplification.

Thermal noise occurs as a result of the constant, random motion of atoms and molecules. The level of thermal noise in any substance is proportional to the *absolute temperature*, which is measured relative to the absence of all heat (*absolute zero*, also known as *zero kelvin* or 0 K). In the everyday environment, temperatures are always high enough to produce plenty of thermal noise in electronic systems.

In any current-carrying medium, the electrons cause noise impulses as they move from atom to atom. This phenomenon is known as *shot effect*, and the resulting noise is called *shot noise*. All the shot noise in a circuit or device is amplified by succeeding stages along with the desired signals. The amount of shot noise that a device produces is roughly proportional to the current that it carries. Low-current solid-state devices, such as the *gallium-arsenide field-effect transistor* (GaAsFET), generate a minimum of shot noise. Devices of this sort, used in the most sensitive parts of an audio system, can help to ensure that the output at the speakers or headset contains the lowest possible level of amplified shot noise.

Electrical noise that enters an audio system through the power supply is called *conducted noise*. In a vehicle, conducted noise goes from the alternator and spark plugs, through the DC power leads, and into the audio equipment. Conducted noise can be minimized by connecting the audio equipment power wires directly to the vehicle's battery terminals. Filtering the non-grounded (usually positive) DC power wire with a circuit, such as the one shown in Fig. 5-19, can minimize alternator whine. *Resistance wiring* in a vehicle ignition system sometimes reduces noise from the spark plugs.

Mechanical vibration can cause unwanted modulation of the oscillators in a hi-fi tuner and also in the circuitry of an audio amplifier. Such unwanted modulation

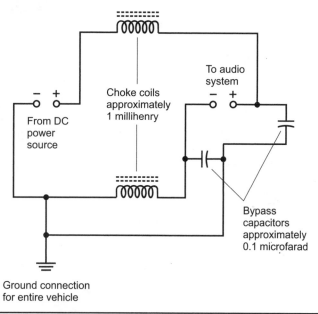

FIGURE 5-19   Choke coils and bypass capacitors can help to minimize electrical noise that can enter your sound system from the power source.

is known as *microphonics*. To minimize this problem, you can mount the entire audio electronic system (except for the speakers) with acoustic padding between the equipment cabinets and the vehicle dashboard or other "solid" object.

Warning!   Never let your sound system's components "float free" in your vehicle. Anchor them down to a fixed surface. In the event you have to suddenly brake, or even worse, if you have a collision or wreck, those objects can turn into deadly projectiles within your vehicle unless they're attached firmly to something.

Fact or Myth?

Some people say that the use of coaxial cable for all "two-wire" inter-component connections in an automotive hi-fi sound system will improve overall performance and minimize the risk of problems with noise and interference. Are they right?

Yes. This rule applies not only for automotive audio systems, but also for all types of audio systems. You must make sure, however, that the outer conductor or "shield" of the cable is connected to a good electrical ground.

CHAPTER 6

Home Entertainment

Hi-fi stereo, TV, and home theater systems can provide you with endless hours of entertainment. You can buy a simple appliance that you take out of the box, plug into a wall outlet or put batteries into it, and "just play it." But sophisticated systems abound, and if you want good sound and video, you'll have to opt for one of them. Let's look at the basics of hi-fi stereo, TV, and Internet access. Then you'll have plenty of ideas as to how you can combine them into a home entertainment system that's ideal for you.

Making the Good Noise

A true hi-fi lover usually assembles a complex set of audio devices into a big system over a period of time, not all at once. In that way, the design ends up best suited to the user's unique needs. Don't hurry as you gather together your "ultimate sound machine." You, as well as the system, must evolve, a process that takes time!

Hi-Fi System Types

The simplest type of home stereo arrangement, called a *compact hi-fi system*, resides in a single cabinet, with an AM/FM radio receiver called a *tuner*, along with a *compact disk* (CD) *player* or *MP3 player*, or both. The speakers can be either internal or external; if they're external, the connecting cables are short. The assets of a compact system are small size, simplicity, and low cost.

More sophisticated hi-fi systems have separate, dedicated equipment cabinets containing components such as:

- An AM/FM tuner
- A CD player
- An MP3 player
- A tape player/recorder for old-fashioned audio tape
- A turntable for old-fashioned vinyl disks
- A satellite radio receiver
- One or more antennas

- An audio mixer
- An amplifier
- A graphic equalizer
- Large, complex speakers
- A headset
- A computer

The individual hardware units in this type of system, known as a *component hi-fi system*, should be interconnected with shielded coaxial cables. A component system costs more than a compact system, but you get better sound fidelity, more audio power, the ability to do more tasks, and the opportunity to tailor the system to your preferences as time goes by.

Some hi-fi manufacturers build all their equipment cabinets to a single, standardized width so that you can install them, one above the other, in a vertical *rack*. A so-called *rack-mounted hi-fi system* saves floor space and gives the system a certain professional look! The rack can be mounted on wheels so that you can easily move the whole system, except for external speakers, from place to place.

Each individual equipment *chassis*, on which all the electronic components are mounted, should be connected to a good electrical ground to minimize hum and noise, and to minimize susceptibility to interference from external sources. If your home has a good three-wire electrical system, the "third slot" in any wall outlet will serve this purpose.

Antennas

In most home hi-fi tuners, the AM antenna is a small coil called a *loopstick* built into the cabinet or mounted on the rear panel. Usually, the FM antenna must be outside and some distance away from the radio. You'll find a connector or pair of terminals for it on the back panel. An FM antenna can consist of a length of TV type *twinlead*, also known as *300-ohm ribbon*, connected to the center of a four-to-six-foot length of wire to form a T-shaped configuration called a *dipole*. You can get ready-made FM dipoles at most good hi-fi stores for a few dollars. You can also use old-fashioned TV *rabbit ears* as the antenna for your FM tuner.

If you live in a fringe reception area and you want good FM radio reception, you'll need an outdoor antenna equipped with lightning protection hardware. Radio Shack stores are a good place to shop for them. Large home appliance outlets often carry them too. For all intents and purposes, an FM outdoor antenna is the equivalent of an old-fashioned outdoor TV antenna. Some people who do not have access to cable TV use their outdoor TV antennas for FM radio reception as well.

As people rely on the Internet more and more for radio and television reception, using traditional receivers less and less, you should expect that antennas will eventually become rare in home entertainment systems, except for the outdoor dish antennas that go along with satellite TV and satellite Internet systems. Unless you're an electronics technician, you'll need to hire a professional to install any dish antenna because they must be precisely aligned with a specific satellite in order to function properly.

Tuner

A typical tuner can receive signals in the standard AM broadcast band (535 to 1605 kHz) and/or the standard FM broadcast band (88 to 108 MHz). Some tuners can also receive satellite radio signals if you have a subscription to a service of that sort. Tuners don't have built-in amplifiers. A tuner can provide enough power to drive a headset, but you'll probably want to add an "outboard" amplifier to provide sufficient power for a pair of speakers.

Modern hi-fi tuners employ *frequency synthesizers* and have *digital displays*. Most tuners have several programmable *memory channels* that allow you to select your favorite stations with a push of a single button, no matter where the stations happen to be in the frequency band. Some tuners also have *seek* and/or *scan* modes that allow the radio to automatically search the band for any station strong enough to come in clearly.

Balance Control

The *balance control* allows adjustment of the relative volume levels of the sounds coming from the left and right channels.

In a basic hi-fi system, the balance control comprises a single rotatable knob connected to a pair of *potentiometers* (variable resistors). When you turn the knob counterclockwise, the left-channel volume increases and the right-channel volume decreases. When you turn the knob clockwise, the right-channel volume increases and the left-channel volume decreases. In more sophisticated sound systems, you can adjust the balance using two independent volume controls, one for the left channel and the other for the right channel.

Proper balance is important in stereo hi-fi. A balance control can compensate for such factors as variations in speaker placement, relative loudness in the channels, and the acoustical characteristics of the room in which the equipment is installed.

Tone Control and Roll-Off

The loudness-versus-frequency characteristics of a hi-fi sound system are adjusted by means of a *tone control*, although the term "tone" is technically a misnomer; a better term would be *frequency-response control*. In its simplest form, a tone control consists of a single rotatable knob or linear-motion sliding control. The counter–clockwise, lower, or left-hand settings of this control result in strong bass and weak treble audio output. The clockwise, upper, or right-hand settings result in weak bass and strong treble. When you set the tone control to mid-position, the audio response of the amplifier is essentially flat, meaning that the bass, midrange, and treble loudness levels are in roughly the same proportions as they were in the original sound recording.

Figure 6-1A shows an arrangement in which a single-potentiometer tone control can be incorporated into the output circuit of an audio amplifier. The amplifier is designed so that its treble output is exaggerated in the absence of any tone control. Potentiometer X *attenuates* (makes weaker) the treble to a variable extent. When you set the potentiometer at zero resistance, the capacitor appears directly across the audio signal path, producing *treble roll-off* (decreasing volume with increasing frequency). As the resistance of the potentiometer increases, the treble roll-off becomes less pronounced. At the mid-point of the potentiometer setting, the treble roll-off caused by the *resistance-capacitance (RC) circuit* cancels out the effect of the exaggerated treble response in the amplifier, and you get a flat frequency response. When you set the potentiometer at its maximum resistance, there's practically no capacitance across the audio signal path. Then the *RC* combination looks like an open circuit (as if neither the potentiometer nor the capacitor were there at all), and the treble response is exaggerated. This type of control causes a general increase in output volume as the resistance of the potentiometer increases; it interacts with the volume control.

A more versatile tone control (Fig. 6-1B) has two capacitors and two potentiometers that you can adjust independently. The amplifier is designed so that it has a flat loudness-versus-frequency output characteristic in the absence of any tone control. The *RC* circuit labeled X produces adjustable treble roll-off, exactly as does the circuit shown at A. When you set potentiometer X at zero resistance, the capacitor appears across the audio signal, causing treble roll-off. As the resistance of the potentiometer increases, the roll-off becomes less pronounced. When you set potentiometer X at its maximum resistance, there's practically no capacitance

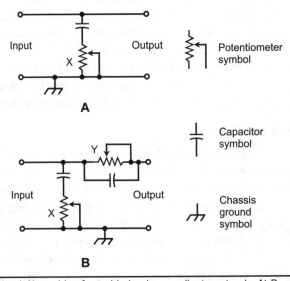

FIGURE 6-1 At A, circuit X provides for treble loudness adjustment only. At B, circuit X allows adjustment of the treble loudness, and circuit Y allows adjustment of the bass loudness.

across the audio signal. Then the *RC* combination appears as an open circuit, and you get no treble roll-off at all. The *RC* circuit labeled Y attenuates the bass to a variable extent. When potentiometer Y is at its maximum resistance, the capacitor across it causes a *bass roll-off* (decreasing volume with decreasing frequency). As the resistance of the potentiometer decreases, the effect of the capacitor diminishes, and the roll-off becomes less pronounced. When you set potentiometer Y at zero resistance, you get essentially no bass roll-off.

In either of the arrangements shown in Fig. 6-1, the optimum values of the potentiometers and capacitors must be found by experimentation, a task normally done by the engineers who design the equipment. Of course, each audio channel needs its own dedicated tone control circuit, so a stereo amplifier will have two tone controls, one for the left-hand channel and the other for the right-hand channel.

Sophisticated tone controls, commonly found in most systems these days, make use of specialized *integrated circuits* (ICs) called *operational amplifiers* or *op amps*. These devices can be "tweaked" to have almost any desired loudness-versus-frequency characteristic.

Did You Know?

In some high-end and professional audio systems, a technology known as *digital signal processing* (DSP) is used to obtain tone control that has enough versatility to satisfy the most demanding "tweak freak."

Audio Mixer

If you connect two or more audio sources directly to the same input jack or terminals of an amplifier, you can't expect good results. Different signal sources, such as a computer, a tuner, and a CD player will almost certainly have different AC resistances, called *impedance values*, for audio signals. When connected together, these impedances appear in parallel, that is, across each other like the rungs in a ladder. This sort of situation can hinder the performance of the sound-generating devices, as well as mess things up at the amplifier input terminals. You'll end up with low system efficiency and poor overall performance. In some cases, one or more of the input devices will fail to produce any sound at all in your amplifier.

Another problem with direct interconnection of multiple sound sources arises from the fact that the signal levels from the devices usually differ. A microphone produces tiny audio-frequency currents, whereas a tuner produces enough to drive a headset or even a small speaker. Connecting the outputs of these devices directly across each other will cause the microphone signal to be obliterated by the signal from the tuner, and the tuner's output audio current might physically damage the microphone.

An *audio mixer* eliminates the problems that you face when you want to connect the outputs of multiple audio devices to a single channel input for an amplifier. The mixer isolates the input impedances from each other, so you don't have to worry about any possible mismatch or "competition" among the source devices. In addition, you can adjust the signal level or *gain* for each device without affecting the behavior of any other device. That way, you won't be surprised by near silence from one device or a gigantic sound blast from another.

Equalizer

An equalizer allows for the adjustment of loudness of audio signals at various frequencies. You learned about this type of device in Chapter 5. Equalizers serve the same function in a home hi-fi system as they do in a mobile system. In general, however, equalizers for home systems are more sophisticated (and expensive) than those for automotive systems. You'll find graphic and parametric equalizers commonly available for home audio systems. After all, you can "fool around with the controls" a lot more easily in a fixed system than you can do in a mobile system because you don't have to worry about causing a motor-vehicle wreck if you get distracted!

Making Sound Loud

In hi-fi systems, an amplifier delivers significant audio power to a set of speakers. An amplifier always has at least one input, but more often there are three or more: one for a CD player, another for a tuner, and still others for auxiliary devices, such as a tape player, turntable, or computer.

How Loud Is Loud?

You don't perceive the loudness of sound in direct proportion to the power contained in the acoustic waves. Instead, your ears and brain sense sound levels according to the logarithm of the actual intensity. Another variable is the *phase* with which waves arrive at your ears. Phase allows you to perceive the direction from which a sound is coming, and it also affects the apparent sound volume as you hear it.

Engineers express sound levels in units called decibels (dB). If you change the volume control on a hi-fi set so that you can just barely tell the difference in the loudness *when you anticipate the change*, then that change equals approximately 1 dB. In acoustic applications, decibels express relative, not absolute, sound power. If you adjust an amplifier's volume control so as to double the actual sound power coming from a set of speakers, then you cause a volume change of +3 dB (3 dB of *gain*). Conversely, if you halve the sound power, you get a volume change of −3 dB (3 dB of *loss*). Increases in sound power go along with positive decibel values, and decreases in sound power go along with negative decibel values.

If you want decibels to mean anything, you need a reference volume level against which you can compare all other sounds. Have you been told that a large electric vacuum cleaner produces 80 dB as heard by the person operating it? This figure is determined with respect to the *threshold of hearing*, which represents the faintest sound that a person with "good ears" can detect in a *quiet room* specially designed to have a minimum of background noise.

Did You Know?

Large stereo systems can produce sound levels in excess of 100 dB throughout a good-sized living room. That's more than loud enough to cause permanent damage to your ears if you listen to it for a long time. An audio amplifier and speaker system cranking out 100 dB makes as much racket as a jackhammer at close range.

Amplifier Linearity

Linearity is the extent to which an amplifier's output *waveform* (the shape of the wave as it would look on the screen of a laboratory oscilloscope) represents a faithful, exact magnification of the input waveform. In hi-fi equipment, all amplifiers must be as *linear* as possible. You can think of audio amplifier linearity as the equivalent of optical precision in a microscope or telescope.

If you connect a *dual-trace oscilloscope* (one that lets you observe two waveforms at the same time) to the input and output terminals of a hi-fi audio amplifier that has good linearity, the output waveform will show up as a vertically magnified duplicate of the input waveform. If you apply the amplifier's input signal to the horizontal scope input and the amplifier's output signal to the vertical scope input, you'll see a straight line on the screen. In an amplifier with poor linearity, the line on the screen will appear "bent" or "kinked," indicating that the output waveform is not a faithful reproduction of the input. In that case, you know that the amplifier produces *distortion*. Sometimes you can't notice distortion simply by listening to the sound, even if that distortion shows up clearly on an oscilloscope.

For Nerds Only

Hi-fi amplifiers are designed to work with input signals up to a certain *peak* (maximum instantaneous) *amplitude*. If the peak input amplitude exceeds the critical level, the amplifier becomes nonlinear, and distortion takes place. In a hi-fi system equipped with *volume-unit meters* (also called VU meters) or *distortion meters*, excessive input causes the meter needles, or bars, to "kick up" into the red range of the scales during audio peaks. You should always operate your amplifiers so that the VU meter readings stay well below the "red zone."

Dynamic Range

When engineers talk about *dynamic range* in a hi-fi audio system, they refer to the difference between the strongest and the weakest output audio signals that the system can produce without objectionable distortion taking place. Dynamic range is usually specified in decibels, by comparing the strongest and weakest signal levels in logarithmic terms. It's a prime consideration in hi-fi recording and reproduction. As the dynamic range specification of an amplifier increases, the sound quality improves for music or programming having a wide range of volume levels. You'll want to compare this specification among different amplifiers when you're shopping.

At low volume levels, the limiting factor in dynamic range is the *background noise* in the system. At high volume levels, the power-handling capability of an audio amplifier limits the dynamic range. If all other factors are equal, you can expect a 200-watt audio system to have greater dynamic range than a 100-watt system. The speaker size is also important. As speakers get physically larger, their ability to handle high power improves, resulting in increased dynamic range for the entire amplifier/speaker arrangement.

Did You Know?

An "overengineered" sound system can be a symptom of *megalomania*, an affliction that commonly overtakes the minds of hard-core audiophiles. But this apparent madness has a method behind it. A huge dynamic range number ensures that occasional extreme audio peaks, however rare, can be reproduced without distortion. If you want true concert-hall-quality sound, you can't put up with any detectable distortion, ever! Of course, megalomania can prove expensive to those who suffer from it, but if money is no object, you'll want to go for the biggest dynamic range number you can find.

Getting Sound Out

No amplifier can deliver sound that's any better than the speakers will allow. Speakers are rated according to the audio power they can handle. It's a good idea to purchase speakers that can tolerate at least twice the audio output power that the amplifier can deliver. Such "overengineering" will ensure that distortion will not occur in the speakers during loud, low-frequency sound bursts. Overengineering in this department will also minimize the risk of physical damage to the speakers that might otherwise result from accidentally *overdriving* them.

Dynamic Speaker

A *dynamic speaker* comprises a coil and magnet that translate electrical current into mechanical vibration, thereby producing sound waves in the air. Figure 6-2 is a functional (but not literal) illustration of a dynamic speaker. A *diaphragm*, which sometimes takes the form of a *speaker cone*, is attached to a coil that can move back and forth rapidly along its axis. If you apply an audio signal to the coil, the variable current in the coil generates a fluctuating magnetic field that produces forces on the coil as it interacts with the permanent magnet's field. These forces cause the coil to move, pushing the diaphragm back and forth to create acoustic waves in the surrounding air.

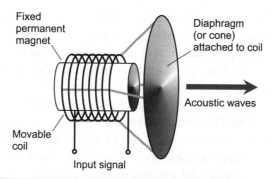

FIGURE 6-2   Functional diagram of a dynamic speaker.

Electrostatic Speaker

An *electrostatic speaker* takes advantage of the forces produced by electric fields rather than magnetic fields. Two large, flat metal plates, one flexible and thin and the other rigid and thick, are placed parallel and close together as shown in the functional diagram of Fig. 6-3. The *blocking capacitor* allows for the connection of a high DC voltage between the speaker plates, but keeps the DC from getting back into the amplifier system and disrupting its performance. (Capacitors block DC but allow AC, such as audio signals, to pass through.) The fluctuating, high AC audio voltage that comes out of the transformer, combined with the DC voltage between the plates, gives rise to a powerful, fluctuating electrostatic field between the plates. That field produces a variable force on the flexible plate with respect to the rigid plate, so that the flexible plate "bows in and out" and causes sound waves to arise in the air.

Speaker Cabinets

Good speakers contain two or three individual "signal-to-sound converters" (technically a form of *electromechanical transducer*) within a single cabinet. The woofer

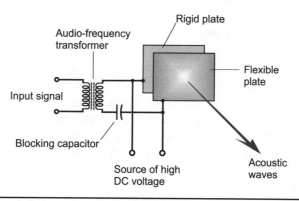

FIGURE 6-3   Functional diagram of an electrostatic speaker.

reproduces the bass (low-frequency) sound. The midrange speaker handles medium and, sometimes, treble (high-frequency) sound. A tweeter is designed especially for enhanced treble reproduction, and most good ones can turn signals at frequencies above the human hearing range into *ultrasound*. The design of the cabinet has a profound effect on the quality of the sound that comes out of the speakers.

Because a speaker cabinet is an enclosed chamber, it will have *resonant frequencies* at which it reinforces the sound waves inside itself, and also *null frequencies* at which it cancels out the sound waves inside itself. These resonant and null effects should be minimized to keep the speakers from producing artificially exaggerated loudness at the resonant frequencies and artificially muted sound at the null frequencies. Speaker manufacturers use a variety of techniques in order to achieve this goal. Foam padding and internal *baffles* (sound reflectors) are common.

Wave Shapes and Reflections

The shape, as well as the frequency, of a sound wave affects the manner in which and the extent to which the acoustic disturbance reflects from various physical objects as it travels through the air from the speakers to your ears. Acoustics engineers must consider the *waveform* (wave shape) when designing sound systems and concert halls. The goal is to make sure that all the musical instruments sound realistic everywhere in the room—or at least, that they get close to that ideal. Computer models and simulations can help with this process, but in the end, trial-and-error experimentation is necessary. Judgment must be made subjectively by the listeners. No matter how well a system works in theory, if the listeners don't like the way it "plays," the finest mathematical models don't mean a thing!

Suppose that you have set up a sound system in your living room, and that, for the particular placement of speakers with respect to your ears, sounds propagate well at 1, 3, and 5 kHz, but poorly at 2, 4, and 6 kHz. This situation will affect the way that the various musical instruments sound. You'll end up with more distortion in the sounds from some instruments than in the sounds from other instruments. Unless all sounds, at all frequencies, reach your ears in the same proportions that they come from the speakers, you won't hear the music as it originally came from the instruments.

Figure 6-4 shows a listener, a speaker, and three baffles as they might be arranged in a large room. The waves X, Y, and Z as they reflect from the baffles, along with the direct-path wave D, add up to something different at the listener's ears for each frequency of sound. The way that the sound waves combine will change as the listener moves around the room. This phenomenon is impossible to prevent entirely; the best you can hope for is to minimize it. That's why even the best engineers find it difficult to design an acoustical room, such as a concert auditorium, that will propagate sound in an optimum way at all frequencies for every listener.

Headsets

A *headset* offers listening privacy, keeps your "big sound" experience from disturbing people around you, and gets rid of sound-wave reflection problems

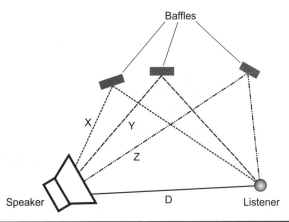

Figure 6-4   Reflected-path sound waves X, Y, and Z combine with the direct-path sound wave D to produce what the listener hears.

inherent in all systems that use speakers. In effect, a headset comprises two small dynamic speakers, one placed directly against (or very close to) each ear.

Two equally expensive headsets can exhibit huge differences in the quality of the sound that they put out, and people will always disagree about what constitutes good sound. A good hi-fi store will have several headsets on display, connected to a sound system so that you can "test listen" to each one and then choose the headset that you like the best.

Television Then and Now

Television (TV) grew popular in the United States during the 1950s, and became firmly established by the end of 1960 after the airing of the debates between John F. Kennedy and Richard M. Nixon in their contest for President of the United States. The first TV sets were built into cabinets that weighed well over 200 pounds and employed vacuum tubes at every stage. The "picture tube," a primitive version of a *cathode-ray-tube* (CRT) display, produced a blurred, grayscale image with a diagonal measure of less than two feet. The signals employed *amplitude modulation* (AM) like the standard radio broadcasts of the same era, and used *analog* (continuously variable) methods to convey both the image and the sound.

Analog versus Digital

Old-fashioned analog television is also known as *fast-scan TV* (FSTV) or *National Television System Committee* (NTSC) TV. In most of the world, broadcasters no longer use this mode; it was pretty much done away with worldwide by 2011. Nevertheless, if you have an old TV set, chances are good that it was designed for analog TV and won't work nowadays unless you get a digital-to-analog converter box.

In analog TV, the individual images, called *frames*, were transmitted at the rate of 30 per second. There were 525 lines of video information per frame. Color NTSC

TV worked by sending three separate monochromatic signals, corresponding to the *primary colors* red, blue, and green. The signals, in effect, were "redscale," "bluescale," and "greenscale." The receiver recombined these signals and displayed the resulting video as a fine, interwoven matrix of red, blue, and green dots.

When viewed from a distance, the image dots in an analog color TV display were too small to be individually discernible, but you could easily see them close-up. Various combinations of red, blue, and green intensities could yield any color that the human eye can perceive. In a good analog color TV set receiving a strong signal, the color quality was excellent, even by today's standards, but the detail left something to be desired.

For Aging Nerds Only

Figure 6-5 is a time-domain graph of a single line in an NTSC video signal, representing 1/525 of a frame. The highest signal level corresponded to the blackest shade, and the lowest signal level corresponded to the lightest shade. Therefore, the FSTV signal was sent "negatively." This convention allowed for *retracing* (moving from the end of one line to the beginning of the next) to synchronize between the transmitter and receiver. A well-defined, strong *blanking pulse* told the receiver when to retrace, and it also shut off the beam while the receiver display was retracing. Even so, weak signals looked "snowy" or "washed-out." Today, with digital systems, weak signals usually fail come in altogether, so you get either a good image or none at all.

The term *high-definition television* (HDTV) refers to any of several methods for getting more detail into a TV picture than could ever be done with NTSC TV. The HDTV mode also offers superior sound quality, making for a more satisfying home TV and home theater experience. High-definition TV is sent in a digital mode; this offers another advantage over analog TV. Digital signals propagate better than analog signals do, they're easier to deal with when they are weak (if they're not good enough, they just go away!), and they can be processed in ways that analog signals would not allow.

The Big Screen Cometh

Television technology has evolved in ways that few people imagined in the 1950s. *Digital modulation*, which conveys both video and audio information in discrete data bits in the same way as computer data is transmitted, has replaced analog modulation. *Solid-state* components, based on semiconductor materials, such as silicon and gallium arsenide, have replaced electron tubes. Color images have supplanted grayscale images. Video detail, called *image resolution*, has improved immeasurably. Today you can enter a department store with a few hundred dollars and come back out with a flat-screen TV set that has a diagonal measure upwards

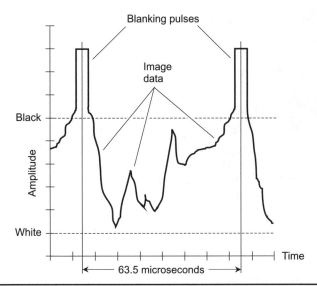

FIGURE 6-5 A single line in an old-fashioned analog TV video frame as it would appear on a lab oscilloscope.

of four feet and offers hi-fi sound that you can interconnect with your home stereo system. Some sets even provide for three-dimensional (3D) image viewing.

When you're in the market for a new TV set, you'll find several types of display. In recent years, display types have converged toward light-emitting diode (LED) technology. A few sets still use old-fashioned CRTs, but most video experts agree that they're not long for this world. Plasma displays, of the sort that department stores sometimes put in prominent places for mass viewing by consumers, exist for use in home video systems, but they, like CRTs, are dated. That leaves LED screens and *liquid-crystal-display* (LCD) screens, which are equivalent for practical purposes. As you shop for a flat-screen TV, you should take the following factors into account:

- Screen size (diagonal measure in inches)
- Weight (in pounds or kilos)
- Resolution (amount of detail in the image)
- Screen surface type (shiny or matte)
- Internet capability (or not)
- Sound capability (hi-fi is standard but not universal)
- 3D capability (or not)
- Brand name (if you have a favorite)
- Cost (if you're on a budget)

No matter how many things you ponder in theory, you should always get a look at your chosen TV set in action before you invest in it. Most stores that sell big-screen TVs have several of them mounted up on a wall, side by side, all showing the same program. You can get a good idea of which set has the brightest and truest colors, and which one produces the clearest image, within a couple of minutes.

Did You Know?

When you look at any TV display, the colors that you see will depend on the lighting environment. If the store has bright fluorescent lamps on the ceiling and your home has small floor or table lamps with shades, you'll perceive different hues at home from the ones you saw in the store. Before you walk out of a retail establishment with a brand new piece of video display hardware like a big-screen TV, make sure that the store's management will let your return it within a few days if you don't like the way it plays in your home.

Pixels

The term *pixel* is a contraction of the words "picture element." A pixel is the smallest unit of visible information in a video image. In a digital color display, each pixel can have any of numerous *hues* (color tints), *saturation* (color richness) levels, and *brightness* (actual brilliance) levels, independently of all the other pixels. Any video display will carry a specification that tells you the number of pixels going vertically; some will tell you both the horizontal and vertical values.

For example, in a display that claims 1080p, there are 1080 individual picture elements in each vertical row on the screen. If a display's specifications tell you that the screen measures 1600 × 900, then you know it has 1600 pixels going horizontally from left to right, and 900 pixels going vertically from top to bottom. (That's *1.44 million* individual video elements, by no means the largest number that you'll ever see!) Even a modest digital TV set can display an image consisting of several hundred thousand pixels in total.

For Nerds (and Everybody Else)

When you get your new big-screen TV installed in your favorite video viewing den and have made sure that it actually works, check out the display with the set switched on but not receiving any signal at all. It should show a uniform gray. Scrutinize the entire display area and look for individual pixels that might show up as either solid white or solid black, indicating a failed LED or LCD element. Don't settle for a set that has any nonfunctional pixels when it's brand new!

Input Ports

A contemporary digital TV set has a connector that goes directly to the coaxial cable through which the incoming video and audio signals arrive. You can recognize this connector by its finely threaded cylindrical form measuring about 1/4 of an inch (a little over 6 millimeters) in diameter. The center of this signal input port has a tiny hole into which the center conductor of the cable goes; the threaded exterior connects to the cable's outer conductor or shield. To make the set play, you will normally connect the signal cable directly to this port, exactly as you would do with any other cable-ready TV set.

The best big-screen TVs have other input signal ports besides the one for conventional TV viewing. Always look for a set that has a *High-Definition Multimedia Interface* (HDMI) port. It will let you connect your big screen set to an up-to-date computer so that you can view programs on the Internet, and also look at homemade videos that you can create using popular devices, such as webcams, camcorders, and Apple iPads. You might also want to have a set that includes a *Video Graphics Array* (VGA) port so that you can connect it to an older computer and use it as a display.

Satellite TV

Until the early 1990s, a satellite television installation required a dish antenna roughly six to 10 feet (two or three meters) in diameter. A few such systems are still in use. The antennas are expensive, they attract attention (sometimes unwanted), and they're subject to damage from ice storms, heavy snows, and high winds. Digitization has changed this situation. In any communications system, digital modes allow the use of smaller receiving antennas, smaller transmitting antennas, and/or lower transmitter power levels. Engineers have managed to get the diameter of the receiving dish down to about two feet or 2/3 of a meter.

The *Radio Corporation of America* (RCA) pioneered digital satellite TV with its so-called *Digital Satellite System* (DSS). The analog signal was changed into digital pulses at the transmitting station using A/D conversion. The digital signal was amplified and sent up to a satellite. The satellite had a *transponder* that received the signal, converted it to a different frequency, and retransmitted it back to the earth. A portable dish picked up the downcoming (or *downlink*) signal. A *tuner* selected the channel. The digital signal was changed back into analog form, suitable for viewing on a conventional FSTV set, by means of *digital-to-analog* (D/A) *conversion*. Although digital satellite TV technology has evolved since the initial days of the RCA DSS, today's systems work in essentially the same way as the original one did.

Figure 6-6 shows two types of dish antennas that you'll find in satellite TV systems. The design at A is by far more common. The signal arrives at a slight angle with respect to the dish axis, reflects from the spherical or paraboloidal metal surface of the dish, and then enters a device called a feed horn, which acts like an "ear for microwaves." The feed horn is connected to a converter that changes the frequency of the signal so that it can travel along the coaxial-cable feed line to the TV equipment inside your house. The whole assembly measures less than a meter wide and a meter long. The design at B, called a *Cassegrain fed-dish* because its geometry resembles that of a *Schmidt-Cassegrain reflecting telescope*, is sometimes found in more remote areas where a larger antenna is necessary. This type of dish may measure more than two meters in diameter. The signal arrives right along the

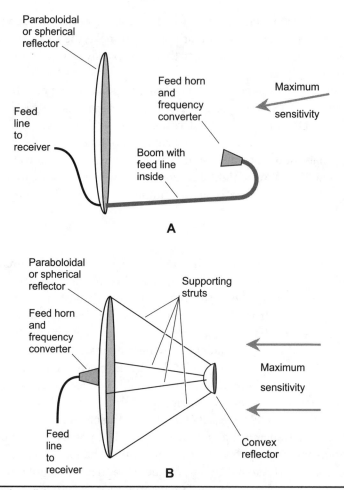

FIGURE 6-6   At A, a dish antenna with conventional feed, commonly used today. At B, a dish antenna with Cassegrain feed, sometimes found in older satellite TV systems.

dish axis, reflects from the spherical or paraboloidal surface, and comes to a focus at a second, smaller reflector. The second reflector causes the incoming microwaves to travel straight back to the center of the dish, where the energy enters the feed-horn-and-frequency-converter assembly through a small hole.

Log-Periodic Antennas

If you live in a place where cable TV service doesn't exist, and if you don't want to pay for a satellite TV system, you can do either of two things: go without TV entirely, or attempt to receive broadcast signals using an outdoor antenna. You can obtain outdoor TV antennas, technically known as log-periodic antennas (LPAs) or log-periodic dipole arrays (LPDAs), along with various other specialized antenna types that can work in either the VHF (very-high-frequency) or UHF (ultra-high-frequency) TV broadcast bands.

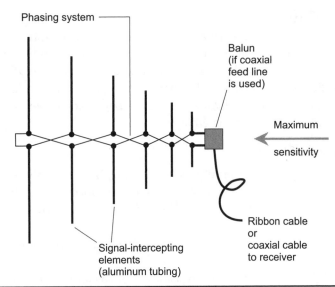

Phasing system

Balun
(if coaxial
feed line
is used)

Maximum

sensitivity

Ribbon cable
or
coaxial cable
to receiver

Signal-intercepting
elements
(aluminum tubing)

FIGURE 6-7   A typical old-fashioned log-periodic antenna for receiving VHF TV. The term "balun" stands for "balanced-to-unbalanced" and refers to a transformer that's needed if you want to feed the antenna with coaxial cable.

The log-periodic antenna receives signals best from a single direction, just as a dish antenna does, but its alignment is a lot less critical than that of a dish. Figure 6-7 shows a functional diagram of a log-periodic antenna. It's designed for reception of signals in the VHF TV band, known as channels 2 through 13 and covering a frequency range of 54 to 216 MHz. You can also use an outdoor TV antenna of this type to receive standard FM broadcast signals in the range of 88 to 108 MHz. The *balun* (an acronym based on the words "balanced" and "unbalanced" is a transformer that you need if you want to use coaxial cable to feed the antenna.

For Nerds Only

A log-periodic antenna consists of several *dipoles*, in effect pairs of "rabbit ears," all of which are connected together with a pair of wires called a *phasing harness*. The shortest dipole is closest to the feed point that connects to the ribbon or cable going down to your receiver. The longest element resides at the back of the antenna, farthest from the feed point.

The Lightning Factor

Lightning presents a hazard to anyone who has a large outdoor antenna installed. The antenna and its feed line can acquire a large *electrostatic charge* during a thundershower. That charge is like the "static" that accumulates on your body when you shuffle around on a carpeted floor on a dry day, but it is far larger. In the case of a nearby lightning strike, the *electromagnetic pulse* (EMP) caused by fast-moving electrons can produce a surge of current in an antenna. A direct lightning hit, should you suffer one, will cause a massive current surge that can start fires and electrocute people. Lightning can also induce dangerous surge currents on utility and telephone lines. Those little old ladies who run around unplugging everything before a heavy thunderstorm are not as stupid or paranoid as you might think! In fact, lightning is more dangerous, statistically, than hurricanes or tornadoes, perhaps because lightning strikes without any warning whatsoever.

The Nature of Lightning

A lightning "bolt," technically called a *stroke*, lasts for only a small fraction of a second, but the extreme current and voltage produces a fantastic amount of power for that brief instant. Four types of lightning exist, as follows:

1. Lightning that occurs within a single cloud (*intracloud*), shown at A in Fig. 6-8
2. Lightning in which electrons flow from a cloud to the earth's surface (*cloud-to-ground*), shown at B in Fig. 6-8

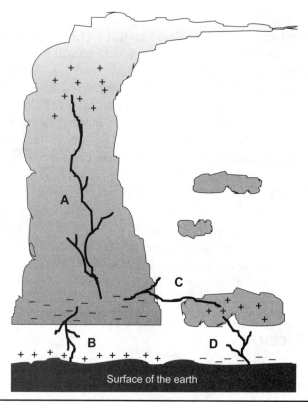

FIGURE 6-8   Four types of lightning stroke: intracloud (A), cloud-to-ground (B), intercloud (C), and ground-to-cloud (D).

3. Lightning that occurs between two clouds (*intercloud*), shown at C in Fig. 6-8
4. Lightning in which the electrons flow from the earth's surface to a cloud (*ground-to-cloud*), shown at D in Fig. 6-8

Cloud-to-ground and ground-to-cloud lightning present the greatest danger to home electronics equipment. Intracloud or intercloud lightning can cause an EMP sufficient to damage sensitive apparatus, even though the lightning "bolt" itself never reaches the surface.

Warning! Whenever you install an outdoor antenna, especially a large one such as a log-periodic that has a considerable amount of exposed metal, you must take the lightning hazard seriously. You should install a lightning arrestor at the point where the antenna's feed line enters the house. In addition, it's a good idea to disconnect the feed line completely from the TV receiving equipment when you aren't actually watching TV. Even then, you can't completely eliminate the danger posed by an outdoor antenna system with regards to lightning. An EMP from a nearby lighting strike can induce a surge of current in your antenna and feed line that can damage your TV. If you suffer a direct hit, it might set your house on fire. For this reason, I don't recommend outdoor TV antennas except for those stubborn few folks who absolutely insist on having one (small satellite dishes being the notable exception because they're relatively safe). For detailed information about protecting your home appliances against the effects of lightning, consult a competent communications engineer. If you have any doubts about the fire safety of an electronic installation, consult your local fire inspector.

Protecting Yourself

You can take the following precautions to minimize the hazard to yourself in and near thundershowers. These measures will not guarantee immunity, however. As an old saying goes, "Lightning has a mind of its own." Sometimes, lightning defies logic and seems to operate outside the laws of physics, so beware!

- Remain indoors, or inside a metal enclosure, such as a car, bus, or train.
- Stay away from windows whether they're open or closed, and whether they have coverings or not.
- If you can't get indoors, find a low-lying spot on the ground, such as a ditch or ravine, and squat down with your feet close together and your head between your legs until the threat has passed.
- Avoid lone trees or other isolated, tall objects, such as utility poles or flagpoles.
- Avoid electric appliances, or electronic equipment that makes use of the utility power lines or that has an outdoor antenna.
- Stay out of the shower or bathtub.
- Avoid swimming pools, either indoors or outdoors.

- If you're on a beach, get into a nearby building or shelter immediately. Do not stay out there on the sand!
- Don't use hard-wired telephone sets. (Cordless sets or cell phones are okay.)
- Don't use computers with external modems connected, or that operate from the AC utility lines. (Things like an iPad with a wireless Internet connection are okay.)

Protecting Hardware

Precautions that minimize the risk of damage to electronic equipment (but can't guarantee absolute immunity), particularly hardware such as TVs with outdoor antennas, include the following.

- Never operate any type of electronic device that has an outdoor antenna when a thundershower is taking place near your location.
- Disconnect all antennas, and ground all feed line conductors, to a good electrical ground other than the utility power-line ground. Leave the lines outside the building and connect them to an earth ground that's at least a few feet away from the building.
- When you're not using the equipment, unplug it from the utility outlet.
- When you're not using the equipment, disconnect and ground all antenna rotator cables and other wiring that leads outdoors.
- Lightning arrestors provide some protection from electrostatic-charge buildup, but they can't offer complete safety, and you shouldn't rely on them for absolute protection.
- *Lightning rods* reduce (but don't eliminate) the chance of a direct hit, and they can help protect your house from fire in case a direct hit does occur, but they should not be used as an excuse to neglect other precautions.
- Power line transient suppressors help to prevent computer "glitches" and can sometimes protect sensitive components in a power supply, but they should not be used as an excuse to neglect the other precautions.
- Connect antenna supporting masts or towers to an earth ground using heavy-gauge wire or braid.
- You'll find other secondary protection devices advertised in electronics-related and radio-related magazines. You can also consult a competent electrician or your local fire inspector, or both.

Getting Wise to the Web

If you live in a location that offers decent cable service, accessing the *Internet* (also called the *Web*) is as simple as running any other appliance. However, if you live in a rural area or a remote wilderness retreat, or if you haven't used the Internet very much in the past but want to start now, you might want to gain a little bit of web-related wisdom so that you can get the most out of the experience.

Choices, Choices!

If you want to get Internet access in your home and you happen to live in a large city, you have several choices, all of them good. Cable TV providers commonly bundle their services so that you can have Internet access (and sometimes landline telephone service as well). Usually, you'll find that cable Internet connections offer the highest data speeds, measured in *megabits per second* (Mbit/s), where a *megabit* represents a million (1,000,000) individual digits of information. Wireless is a little slower than cable, followed by satellite service and finally dialup, which is excruciatingly slow for today's Internet applications. If you live in a truly progressive town, you might be able to get fiber-optic service directly into your house; this mode is faster still, in some cases offering speeds in excess of a *gigabit per second* (Gbit/s) where a gigabit represents a billion (1,000,000,000) digits of data!

Did You Know?

Kansas City, Missouri and its sister town in Kansas worked out a deal with the Internet giant Google in 2012 to provide fiber-optic Internet service all the way down to residential homes and small businesses. More metro areas will doubtless follow their lead in years to come. Check out the availability of so-called *fiber to the home* in your locale. If it's available and you can afford it, I recommend that you subscribe straightaway. (It'll spoil you in a hurry.)

Wireless service can provide good connection speeds in large cities, but it's rarely as fast as cable. In order to use a wireless Internet service, you'll probably have to get it from a cell phone provider. Most cell phone providers these days offer bundle deals in which you can get a phone set along with wireless Internet service, but some of these plans are quite expensive, and some service providers limit the amount of data that you can upload and download per month. If you want to use the Internet only for *electronic mail* (e-mail) or casual Web browsing, this type of service might suffice for you. However, if you intend to download lots of movies or watch a lot of videos, or if you expect to use the Internet for online gaming, you'll do better with cable (or ideally, fiber-optic) service.

In rural locations too far removed from cell phone towers to get good wireless Internet service (or if the local cell phone provider doesn't include wireless service in any of their packages), you'll probably want to opt for a satellite Internet connection. This mode works like satellite TV, and uses a dish antenna similar to the ones you see in satellite TV installations. The important difference is the fact that with a satellite Internet connection, your dish antenna doesn't merely receive the signals. It transmits signals too, all the way up to a satellite that hovers thousands of miles above the earth! For that reason, you'll need to have a professional technician install your system and align the dish because the antenna alignment is more critical for satellite Internet than it is for TV reception.

The mode of last resort is a dialup connection, in which you use a landline telephone system to access the Internet. It's so slow that most people will find it useless for Web browsing these days; video and streaming audio are out of the question. However, if it's all you can afford, it's better than nothing. Be forewarned, however: Some telecommunications companies have begun to talk about getting rid of landline telephone service altogether. When and if that day comes, you won't be able to get dialup Internet service at all.

Modems

The term *modem* is an acronym that derives from the first letters of the words "modulator" and "demodulator." In technical terms, those words describe precisely what the thing does. It *modulates* (or encodes) signals going out from your computer into the wilds of the Web, and it *demodulates* (or decodes) signals coming in from the Web to your computer. Several different types of modems exist, and the type that you get will depend on how you want to connect to the Internet. A modem can link a computer to the same cable system as you get your TV service from. Some modems are designed to connect directly to a network of optical fibers. Still others contain a small radio transceiver for wireless or satellite access. The most primitive modems work with a telephone landline to get you a dialup connection.

Most new notebook and tablet computers come equipped with internal wireless modems, so you don't have to think about them at all in order to use them. In fact, with some computers, you have to actively disable the internal wireless modem if you want to make sure that your computer doesn't automatically connect to the Internet without your knowledge! To take advantage of the Internet with a wireless-equipped computer or tablet device, you can go to a so-called *wireless hotspot*, follow the instructions provided with your device, and get online. Most public libraries, and a lot of restaurants and bars, provide wireless hotspots. So do hotels, airports, bus terminals, and some retail establishments.

If you want to use the Internet with a cable or satellite system, you'll need an external modem. Your cable provider will probably be willing to supply you with one as part of your monthly TV-and-Internet subscription. They'll sell or rent you a cable modem. It's a box roughly the size and shape of a paperback novel, and it sits or stands on your desk next to your computer. Figure 6-9 shows a cable modem on the author's desk, sitting right underneath the base unit for a cordless phone set. (The upright box to the left of the phone and modem is a *wireless router*, which we'll talk about in a moment.)

Ever since the birth of the Internet, modems have done the same thing: convert signals from a form that travels over a communications medium to a form that your computer can "understand," and vice-versa. They do it faster now than they did in the early years; that's all. Your computer works with *binary digital* signals that occur in *bits* (technically *binary digits*) that can represent either the number 1 or the number 0, but nothing else. These signals are rapidly fluctuating direct currents. In order for binary (two-state) digital information to go over a communications system, the data

Figure 6-9 A cable modem, a cell-phone base unit, and a wireless router. Note the cup with the writing instruments for size comparison.

must be converted to some form of AC signal. That signal can be an electric current in a cable made out of ordinary copper wire, a light beam or infrared beam in an optical fiber, or a radio wave through the atmosphere or outer space.

Routers

A *router* is a device that allows you to access a single Internet connection with more than one computer, although you can use a router if you have only one computer. Routers come in two types: *hard-wired* and *wireless*. To use a router, you plug it into your modem in place of a computer, activate the router according to the instruction manual, and then access the Internet from your computers through the router and the modem combined. Routers will work with cable or satellite Internet connections. You can use a router with a wireless Internet service, too, although there's little reason to do that unless you want the router to act as a *firewall* between your computers and the Internet.

Quick Question, Quick Answer

- What's a firewall? What does it do?
- A firewall is a system that helps protect your computers (and, increasingly, e-book readers and cell phones) against hackers, who might try to gain access to your devices and use them for nefarious purposes. All good computers make use of *software firewalls*, which are programs that either come standard with the machine or that you can install. A router offers extra protection, over and above a software firewall, because it keeps nosey hackers from easily "seeing" your computer by remote control over the Internet. It's a little like shutting the window blinds at night to keep peepers from looking in at you! For this reason, you will sometimes hear that a router can serve as a *hardware firewall*.

With a hard-wired router, you'll need to connect every computer to it using a cord called an *Ethernet cable*, which looks like a telephone landline cord but has more wires. You'll also need an Ethernet cable to hook your router up to your modem. Hard-wired routers aren't convenient if you want to use computers all over the house, and especially if you want to move them around freely. Hard-wired routers are out of the question if you want to use wireless-equipped tablet devices or e-book readers, such as the iPad, Kindle Fire, or Nook Tablet. These devices lack Ethernet ports (jacks for Ethernet cables), but an increasing number of them do have tiny wireless modems that you can use with any wireless router in a home or business.

Fact or Myth?

If anyone tells you that *all* e-book readers will work with wireless routers or wireless Internet connections, don't believe it! Some of them require you to use a computer to download your e-books, and then transfer them to the reader itself, using a special connecting cable. If you are interested in buying an e-book reader and you want it to have wireless capability, make sure that you read the specifications carefully to avoid disappointment.

Because of their convenience and ease of installation, most people use wireless routers when they want to assemble and use in-home *local area networks* (called LANs by techies). That way, they can have multiple computers and tablet devices accessing the Internet all over the house. Figure 6-10 is a block diagram of a wireless LAN serving two computers and two tablet devices. (You can add more devices simply by bringing them within range of the router and switching them on.) The little triangle symbols represent antennas, which are usually inside the devices so you can't see them. The dotted gray lines represent radio waves that travel between your wireless router and the individual devices. Most wireless routers have a maximum range of 100 feet or so, and although that might seem like a limitation, it's actually a good thing. If wireless routers had a much greater range than that, it would increase the risk that unauthorized people might get into your home LAN, especially if you live in a large city or in an apartment or condominium building. Wireless routers require passwords for access, but that security provision can't guarantee that some smart kid won't hack into your LAN anyway.

Did You Know?

Although you can use more than one device at a time in a wireless network, you'll notice that they'll slow down if you have too many of them going at once. That's because the router must allocate the data from the modem among the devices like slicing up a pie. The modem gets only a certain amount data from the service provider; adding a router can't make your modem "suck any more data" from the Internet.

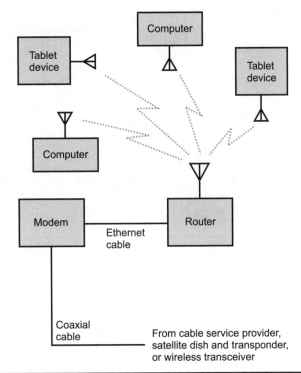

FIGURE 6-10 A home computer network with a modem, wireless router, two computers, and two tablet devices. The little triangle symbols represent the internal antennas in the router, computers, and tablets.

Satellite Anomalies

If you enjoy exotic electronics, and if you live in a remote area where neither cable nor wireless services exist, then you'll doubtless consider having a satellite Internet system installed. Several vendors offer Internet access along with satellite TV service. In the United States, *DirecTV* and *Dish Network* are both popular. You can also get a stand-alone satellite Internet connection through a service, such as *Starband*, *Wild Blue*, or *Hughesnet*. Although a good satellite Internet connection can save you from the drudgery of dialup, you will have to contend with certain technical anomalies. Once in awhile, a disgruntled satellite Internet subscriber lets their service contract expire and goes back to dialup because they could not deal with the inherent limitations of satellite Internet service.

A dedicated satellite Internet dish antenna resembles the "hybrid" TV-and-Internet antenna shown in Fig. 6-11. You'll need to have a professional installer mount and align any satellite Internet dish, whether or not it includes other services, such as TV (or even satellite telephone, which some people rely on in extremely remote places), so that it points exactly at the "bird" that it's meant to reach, and so that it'll stay pointed the right way in case of a windstorm, heavy snowstorm, or other disturbance. Even if you get a perfect installation done, however, you should

Figure 6-11   A typical dish antenna equipped for satellite TV reception and satellite Internet access. It measures about two feet (60 to 70 centimeters) in diameter.

not expect your satellite Internet connection to perform as well as a good cable or wireless system does.

Satellites can contain only a limited amount of electronic memory, processing power, and other resources to give users a fast connection. All the signals from all the subscribers within range must pass through that single satellite. Some services have two or three different satellites, but that's still quite a limitation compared with the hundreds of *nodes* (branch points) in the terrestrial communications network. It's a bottleneck that can get jammed up if every subscriber tries to use the service at once. The provider makes a "deal with fate" in the hope that a situation of that sort won't happen, but if more subscribers sign up and use the service than the vendor originally expected to have, slowdowns and shutdowns can occur in the event of an extreme demand for the service, such as can happen during a natural disaster or national emergency.

The satellite to which your system is assigned will probably follow a *geostationary orbit*, also called a *parking orbit* because the satellite remains fixed in the sky, as "seen" from any point on the surface, for 24 hours a day and seven days a week. In order for the satellite to stay in that special type of orbit, it must fly directly over the earth's equator, going from west to east at an altitude of about 22,500 miles (36,000 kilometers). That way, it revolves around the earth at exactly the same speed, and in the same direction, as the earth rotates on its axis, and always stays directly above the same point on the earth's equator.

When you make a request for a Web page by clicking on a link, tapping on a touch screen, or making whatever other maneuver is required, a signal travels from your dish up to the satellite, then back down to the earth's Internet system to find the server (computer) where the content resides. That server then sends the data back up to the satellite, which, in turn, sends the final bits of digital information to your system. All in all, the signals make two complete round trips between the earth and the satellite for a total of more than 90,000 miles (145,000 kilometers). The radio waves alone take half a second to go that far, so the best possible return time you can expect is half a second. In real life, it's usually more like a full second because systems on the ground impose an additional delay. That long delay, called *latency*, will make it impossible for you to use VoIP (technically known as *Voice over Internet Protocol*, but often called Internet telephone) with the ease and convenience you'll have come to expect if you've done it over a cable connection. Internet gaming will also suffer because of the latency.

As if the limitations imposed by the laws of physics weren't enough, you'll also have to deal with the rules and regulations imposed by the provider. In order to guarantee reliable service for all users, they put a limit on the amount of data that each individual user can receive from (*download*) or transmit to (*upload*) the satellite per 24-hour period. If you use the Internet only for e-mail and text-article viewing, you might not bump up against that limit. But if you download a lot of movies or watch long videos online, you should expect to hit the so-called *fair access policy* (FAP) limit. In that case, the service provider will slow your connection down for awhile, in order to keep you from "hogging the bandwidth" and compromising the performance of all the other users.

Other technical glitches can occasionally get into a satellite Internet system. One of the more interesting effects occurs during the short and infrequent time intervals when the sun passes behind the satellite. These periods last only a few minutes, and they occur only during the middle of March and the end of September; but when the satellite and the sun align as your dish "sees" them, the sun's *radio noise* can overcome the satellite's signal and wipe out your Internet access momentarily. Snow or ice on your dish, extreme temperatures (over 100°F or 38°C), exceptionally heavy rain or snow showers, sandstorms, and dust storms can also disrupt a satellite Internet connection.

For Nerds Only

If you can live with all of the quirks, limitations, and occasional frustrations imposed by the nature of the system; if you have no other good alternative for connection; if you like "techie" stuff; and if you don't take the Internet too seriously, then you can have a lot of fun with a satellite Internet connection. When I had one in the Nevada desert, I used to go out and brush snow off the dish in the winter, and place a wet rag over the transponder on hot summer afternoons to keep it from overheating and shutting down. (I took a photo of that techie's version of a "dish rag" one day in the summer of 2001; I wish I had kept that image so I could reproduce it here and give you a laugh.) On a couple of extreme scorcher days, I aimed a running water hose nozzle at the dish to keep my Internet connection working! Every now and then I'd aim the hose at my landlord's dogs, too, who appreciated the cool bath even more than the satellite dish did.

CHAPTER 7

The Wireless Jungle

The proliferation of wireless devices in recent years has not been accompanied by the level of radio-frequency (RF) *spectrum management* (a fancy term for technological law and order) necessary to ensure compatibility and freedom from mutual interference. Your electric meter might mess up your wireless router. A compact fluorescent lamp (CFL) in a garage-door opener's motor box can interfere with the link between the remote and the motor. A cordless phone set can make an FM radio buzz when the two devices operate near each other. The list goes on.

Dirty Electricity

The following experiment is easy to do, even though the theory gets sophisticated. You'll need a computer with an Internet connection. You'll also need a 12-foot, two-wire audio cord with a 1/8-inch monaural phone plug on one end and spade terminals on the other end (Radio Shack part number 42-2454, for example). If you can't find that component, you can connect a 12-foot length of two-wire speaker cable to a 1/8-inch monaural phone plug.

Electromagnetic Fields

The AC electricity in the utility grid produces obvious effects on appliances: glowing lamps, blowing fans, and chattering television sets. This AC also produces *electromagnetic* (EM) *fields* that aren't apparent to the casual observer. The presence of this EM energy causes tiny currents to flow or circulate in any object that conducts electricity, such as a wire, a metal rain gutter, the metal handle of your lawn mower, and your body. You can easily put together a device that will detect these fields and produce a graphic display of their characteristics.

A current-carrying wire is always surrounded by theoretical *electric flux lines* and *magnetic flux lines*. Around a straight span of wire, the electric or E flux lines run parallel to the wire, and the magnetic or M flux lines encircle the wire, as shown in Fig. 7-1A. If the wire carries constant DC, the electric and magnetic fields are *static*, meaning that the E and M fields stay the same all the time. If the wire carries AC or pulsating DC, the fields fluctuate. The varying E field gives rise to a changing

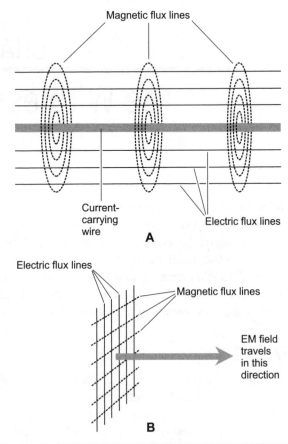

FIGURE 7-1   At A, the electric (E) and magnetic (M) lines of flux around a straight, current-carrying wire. At B, the flux lines far away from a current-carrying wire.

M field, which in turn generates another varying E field. As the E and M fields regenerate each other, a "hybrid field" (the EM field) travels away from the wire, perpendicular to both sets of flux lines at every point in space, as shown in Fig. 7-1B.

All EM fields display three independent properties: *amplitude*, *wavelength*, and *frequency*. The amplitude is the intensity or strength of the field. The frequency is the number of full AC or pulsating DC cycles per second. The wavelength is the distance in space between identical points on adjacent waves. At 60 Hz, the AC utility frequency in the United States, EM waves measure 5000 kilometers (approximately 3100 miles) long in *free space* (air or a vacuum).

A Cool Little Program

You can use your computer to "look at" the EM fields that drench the space all around you. A simple freeware program called *DigiPan*, available on the Internet, can provide a real-time, moving graphical display of EM field components at frequencies ranging from 0 Hz (that is, DC) up to around 5500 Hz. Here's the website: www.digipan.net.

DigiPan shows the frequency along a horizontal axis, while time is portrayed as downward movement all across the whole display. Figure 7-2 shows this scheme. The relative intensity at each frequency appears as a color. You can adjust the colors to suit your taste. If there's no energy at any frequency in the program's range, the display is black. If there's a little bit of energy at a particular frequency, you'll see a thin, vertical blue line creeping straight downward, if you leave the program set for its default color scheme. If there's a moderate amount of energy, the line turns yellow. If there's a lot of energy, the line becomes orange or red. The developers and users of DigiPan and similar programs have coined the term *waterfall* for the display because of its appearance when signals exist at numerous frequencies.

Did You Know?

The DigiPan program was written for digital communication in *phase-shift keying* (PSK), a mode that's popular among amateur radio operators. DigiPan can also function as a very-low-frequency *spectrum monitor*, showing the presence of AC-induced EM fields, not only at 60 Hz (which you should expect) but also at many other frequencies (which you might not expect until you see the evidence). If you have a good Internet connection, DigiPan will download and install in a minute or two.

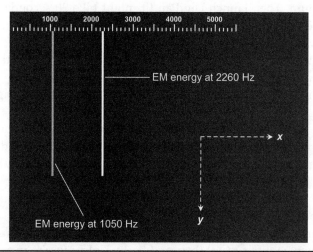

FIGURE 7-2   The DigiPan display system. The horizontal axis portrays frequency. Time "flows" downward across the whole screen. Signals show up as vertical lines. This drawing shows two hypothetical examples.

For Nerds Only

DigiPan isn't the only program you can use to display dirty electricity and other EM spectral effects. Another program, called HamScope, works well too. Go into your favorite Internet search engine and enter the single word "HamScope" to find download sites. This program, like DigiPan, has a small "footprint" on a personal computer, and will download fast. As for learning to use it, you can do as I did: Follow the COEISASWH paradigm (click on everything in sight and see what happens). Different computers will "play" these programs differently, so beware!

The Hardware

To observe the EM energy on your computer, you'll need an antenna. Cut off the U-shaped spade lugs from the audio cord with a scissors or diagonal cutter. Separate the wires by pulling them apart along the entire length of the cord, so that you get a 1/8-inch monaural phone plug with two 12-foot wires attached.

Insert the phone plug into the *microphone* input of your computer. Arrange the two 12-foot wires so that they run in different directions from the phone plug. You can let the wires lie anywhere, as long as you don't trip over them! This arrangement will make the audio cord behave as a *dipole antenna* to pick up EM energy.

Open the audio control program on your computer. If you see a microphone input volume or sensitivity control, set it to maximum. Set your computer to work with an external microphone, not the internal one. If your audio program has a "noise reduction" feature, turn it off. Set the microphone gain (or input sensitivity) to maximum. Then launch DigiPan (it might take 30 seconds or so to load) and follow these three steps, in order.

1. Click on "Options" in the menu bar and uncheck everything except "Rx."
2. Click on "mode" in the menu bar and select "BPSK31."
3. Click on "View" in the menu bar and uncheck everything.

Once you've carried out these steps, the upper part of your computer display should show a jumble of text characters on a white background. The lower part of the screen should be black with a graduated scale at the top, showing numerals 1000, 2000, 3000, and so on. Using your mouse, place the pointer on the upper border of the black region and drag that border upward until the white region with the distracting text vanishes.

If things work correctly, you should have a real-time panoramic display of EM energy from zero to several thousand hertz. Unless you're in a remote location far away from the utility grid, you should see vertical lines of various colors. These lines represent EM energy components at specific frequencies. You can read the frequencies from the graduated scale at the top of the screen. Do you notice a pattern?

Harmonics

A pure AC sine wave appears as a single *pip* or vertical line on the display of a spectrum monitor (Fig. 7-3A). This pip means that all of the energy in the wave is concentrated at one frequency, known as the *fundamental frequency*. But many, if not most, AC utility waves contain *harmonic* energy along with the energy at the fundamental frequency.

A *harmonic frequency* is a whole-number multiple of the fundamental frequency. For example, if 60 Hz is the fundamental frequency, then harmonics can exist at 120 Hz, 180 Hz, 240 Hz, and so on. The 120 Hz wave is at the *second harmonic*, the 180 Hz wave is at the *third harmonic*, the 240 Hz wave is at the *fourth harmonic*, and so on. In general, if a wave has a frequency equal to *n* times the fundamental where *n* is some whole number, then that wave is called the *nth harmonic*. (The fundamental is the first harmonic by definition.) In Fig. 7-3B, a wave is shown along with its second, third, and fourth harmonics, as the entire "signal" would appear on a spectrum monitor.

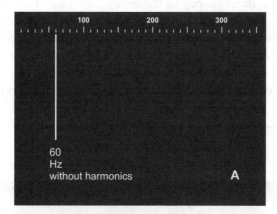

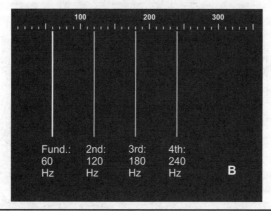

FIGURE 7-3   At A, a spectral diagram of pure, 60-Hz EM energy. At B, a spectral diagram of 60-Hz energy with significant components at the second, third, and fourth harmonic frequencies.

When you look at the EM spectrum display from zero to several thousand hertz using DigiPan, you'll see that utility AC energy contains not only the 60-Hz fundamental, but *many* harmonics. When I saw how much energy exists at the harmonic frequencies in and around my house, my amazement knew no bounds. I had suspected "dirt in the ether," but not *that* much! Figure 7-4 shows my own DigiPan display of dirty electricity from 60 Hz to more than 4000 Hz. Each vertical trace represents an EM signal at a specific frequency. If the electricity were "perfectly clean," you'd see only one bright vertical trace at the extreme left end of the display.

Try This!

Place a plug-in type vacuum cleaner near your EM pickup antenna. Switch the appliance on while watching the DigiPan waterfall. When the motor first starts up, do curves suddenly appear on the display, veering to the right and then straightening out as vertical lines? Those contours indicate energy components that increase in frequency as the motor "revs up" to its operating speed and maintain constant frequencies thereafter. When the motor loses power, do the motor's vertical lines curve back toward the left before they vanish? Those curves indicate falling frequencies as the motor slows down. Try the same tests with a hair dryer, an electric can opener, or any other appliance that plugs into a wall outlet and contains an electric motor. Which types of appliances are the "noisiest"? Which are "quietest"?

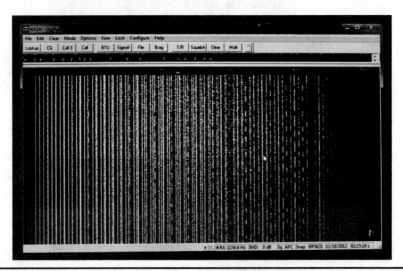

FIGURE 7-4   Dirty electricity at the author's home as viewed on DigiPan.

Fact or Myth?

If you conduct some serious Internet-based research into this subject, you'll come across a lot of sites that warn about health hazards posed by dirty electricity. Some sites will give you case histories, horror stories, and wild tales about illnesses, pains, and cancer, and attribute all of the trouble to dirty electricity. How serious is the danger, really? As a former radio-frequency (RF) engineer and antenna specialist, the best answer I can give you is "I don't know." I suspect, however, that if the "dirt" in dirty electricity has adverse health effects, the "clean" part, which produces far stronger EM fields, probably does as well, and to a far greater extent. So any efforts to "clean up" electricity, in the hopes of getting rid of its potential ill effects, are probably futile.

Cordless Phones

Nowadays, most landline-connected telephone sets are of the cordless type. Rare indeed is the old table-top or wall-mounted phone set with a coiled cord between the receiver and the base unit! Cordless phones offer convenience; that's their main asset. But they also have certain limitations and shortcomings.

Did You Know?

Cordless phones differ from cell phones in some important ways. A cordless phone requires a base unit in your home or business, connected to a cable that runs to a telephone switch (the landline). The handset must normally be within about 100 feet (30 meters) of the base unit. Cell phones have far greater range and do not need a base unit or landline. Cordless phones will "die" if your electric power goes out. Cell phones will keep working unless the power outage is so widespread that it affects the cellular system as well as your local neighborhood.

How They Work

The wireless link in a cordless phone set uses a digital encoding system designed to optimize the sound quality, minimize the effects of interference, and make it difficult for people to eavesdrop on your conversations. That hasn't always been the case.

In the early days of cordless phone technology (the 1980s), the handsets operated at radio frequencies around 1.7 MHz, immediately above the standard AM broadcast band. Power-line interference, poor audio quality, and eavesdropping took place as a general rule! If you had an all-wave or "shortwave" radio receiver, you could tune it to those frequencies and, if you lived in a populated area, hear two or three conversations at any given time! You might even access someone else's line with your handset, or have someone else take advantage of your line.

If you buy a cordless phone set today, it will probably function at 1.9 gigahertz (GHz), where 1 GHz equals 1000 MHz or 1,000,000,000 Hz. Some of them work at 2.4 GHz or 5.8 GHz. These signal waves are only a few inches long as they travel through the air between the handset and the base unit. Instead of a whip antenna a couple of feet long, the handset has a tiny "stub" antenna or an internal antenna. Instead of conventional AM, which the earliest cordless phones used, the new ones employ a technology called *digital spread spectrum* (DSS), which makes it almost impossible for anyone to eavesdrop on your conversations or gain access to your phone line without your consent.

For Nerds Only

In *spread-spectrum communications* modes including DSS, the signal frequency doesn't stay constant like it does in a normal radio system. Instead, the frequency hops or sweeps over a defined range called a *band*. The transmitter frequency varies according to a specific, encoded, repeating pattern. No receiver can hear the signal unless it "knows the code" and acts on it correctly. Exotic as it sounds, spread spectrum is far from new. Among the first people to dream it up were the composer George Antheil and the actress Hedy Lamarr, who received a U.S. patent for a spread-spectrum process in 1942!

Woes and Resolutions

Despite all the advanced technology that cordless phones use (or maybe because of it), you can expect some glitches with their operation. Here's a short list of problems and suggestions for addressing them.

- *Bad battery*—Weak or dead batteries can pose a problem with cordless phones more than a couple of years old. The rechargeable lithium-ion batteries in the handset normally last three to five years, and then you'll have to replace them.
- *They need utility power*—If you experience a power failure, such as might happen during and after a storm, the base unit can't transfer the signals between the landline and the handset(s), so your cordless phone won't work unless you have a backup power source such as an emergency generator.
- *Poor range*—In buildings with a lot of concrete and steel, you shouldn't expect to get the advertised working range between the base unit and the handset(s). If you place the base unit in a subterranean location such as a cellar, the handsets probably won't work very well outside the house.
- *Whooshing and fading*—When you get near the edge of a handset's useful working range, or if signals happen to conflict with each other as they bounce around among the electrical wires and other metal objects in your

house, you'll hear hissing or whooshing sounds, and might even get completely disconnected without warning. If this problem occurs, move closer to the base unit and stay in one place and position.

- *Interference from other devices*—As the number of wireless devices in our everyday lives keeps increasing, the risk of them interfering with each other grows as well. It can be difficult to figure out where the source of *electromagnetic interference* (EMI) lies, but if you happen to bring two conflicting devices within a few inches of each other, you'll get a good clue.
- *Bad audio*—Some cordless handsets simply sound terrible. The only thing you can do in a case like that, other than endure the problem, is return the phone system for an exchange to another brand, or for a refund of your money.
- *Inadvertent contact with buttons*—In the old days, you could hold a phone handset against your ear and shoulder, freeing up both hands. Not now! If you try that with a cordless handset that "bristles with buttons," chances are good that your chin or cheek will actuate one of them, and you'll hear tones or beeps that seem to come out of nowhere. You might even accidentally hang up on the person you're talking with.

Cell Phones

Wireless telephone sets operate in a specialized communications system called *cellular*. Originally, the cellular communications network served mainly traveling business people. Nowadays, most folks regard *cell phones* as necessities, and most cell-phone sets have extra features, such as text messaging programs, Web browsers, video displays, and built-in cameras.

How They Work

A cell phone looks like a hybrid between a cordless telephone handset and a "walkie-talkie," but smaller. Some cell phones have dimensions so tiny that an unsuspecting person might mistake them for packs of chewing gum. A cell-phone unit contains a radio transmitter and receiver combination called a *transceiver*. Transmission and reception take place on different frequencies so that you can talk and listen at the same time, and easily interrupt the other party, if necessary, a communications capability known as *full duplex*.

In an ideal cellular network, every phone set always lies within range of at least one base station (also called a *repeater*), which picks up transmissions from the portable units and retransmits the signals to the telephone network, to the Internet, and to other portable units. A so-called *cell* encompasses the region of coverage for any particular repeater, also known as a *base station*.

When a cell phone operates in motion, say while you ride in a car or on a boat, the set can move around in the network, as shown in Fig. 7-5. The dashed curve represents a hypothetical vehicle path. Base stations (dots) transfer access to the cell phone among themselves, a process called *handoff*. The hexagons show the limits of

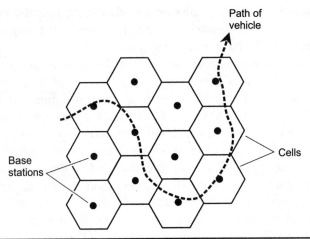

FIGURE 7-5   In an ideal cellular system, a moving cell-phone set (dashed line) always remains within range of at least one base station.

the transmission/reception range (or *cell*) for each base station. All the base stations are connected to the regional telephone system, which, in turn, goes to the major telephone networks. Therefore, from a cell phone on a ranch near Bozeman, Montana (for example), you can place calls to, or receive calls from, almost anyone else in the world.

Cellular connections sometimes suffer from connection problems when signals transfer from one repeater to another. A technology called *code-division multiple access* (CDMA) reduces the prevalence of this problem compared to the early years of cellular technology. In CDMA, the repeater coverage zones might overlap, but signals don't interfere with each other because every phone set possesses a unique signal code. Rather than abruptly switching from one base-station zone to the next, the signal goes through a region in which it flows through more than one base station at a time. This *make-before-break* scheme helps to mitigate cell-transfer trouble.

Did You Know?

Call breakup and dropped calls can occur even in a well-designed and constructed cell-phone network when you make or receive calls from a physical location that suffers from poor reception. In a digital communications network, signals don't fade in and out as they do with analog systems. Instead, the signals appear and disappear, sometimes off-again and on-again in a maddening flurry. Who among us has not experienced this phenomenon while using a cell phone? In some cities, you'll commonly see shattered cell-phone sets lying in gutters or parking lots—victims of *chucking*, where a furious user has hurled a phone set to the pavement with major-league speed. (I don't recommend or condone this practice.)

Woes and Resolutions

As cell phones grow increasingly sophisticated, they can do more and more things. The earliest cell phones were merely glorified "walkie-talkies." Today's *smartphones* allow you to send and receive text messages, browse the Internet, exchange e-mails, create and view photos and videos, manage bank accounts, compare prices in department stores, pay for items at store checkouts, and much more. Along with the versatility and convenience comes a downside: A lot of things can go wrong with these devices. Let's look at the most common complaints.

- Connections break up, or completely disconnect, whenever a cell phone set is not well within the range of at least one repeater, or in rare instances where repeaters conflict. Breakup can also take place if the user moves from a location with a "clear shot" to the nearest repeater (such as an open field) to a location with abundant obstructions (such as a valley or a city street among massive buildings).
- You'll need to keep the battery in good working order and keep it charged, especially if you plan to use the cell phone while traveling. You'll want it to work in a roadside emergency. It's easy to forget this detail, so make it a habit! Also, as with most other battery-powered devices, the battery in a cell phone lasts only three to five years. When the battery dies, you'll have to replace it or upgrade to a new phone.
- Cell phones should not be used while plugged into a charging unit. A few years ago, someone told me a story about a person who got killed when a charging phone exploded in his face! I don't know if the storyteller was "pulling my leg" or not. Lithium batteries don't explode very often, but *they can blow up*, especially if a fully or almost fully charged one gets shorted out.
- Cell phones are somewhat more vulnerable than cordless phones to conflicts with other wireless devices, mainly because cell phones have greater range than cordless handsets do. You'll do best to avoid using cell phones too close to wireless computer peripherals, smart utility meters, active radio transmitters, and sensitive radio receivers.
- Bad audio quality can occur with cordless phones. You can't do much about this problem except get a different phone.
- Depending on where you plan to use your phone, you might need a unit that offers exceptionally loud output audio. Make sure, before you buy a phone set, that it produces enough sound, and in a clear enough state, to satisfy you.
- Some older cell phones have a problem with overly sensitive buttons causing unwanted responses, the same sort of thing that can happen with cordless phones. Newer cell phones have few (or no) exposed buttons that can inadvertently contact your face and cause an unintended reaction.
- Some cell phones "obey" your fingertips better than others do. It's the same sort of variability that you know about if you've used tablet computers and

electronic book (e-book) readers. Before you buy a cell phone set, try to use it in real time. Can you easily send texts with it? Does it accurately register numbers and characters as you enter them in?

- The display might wash out in sunlight, not be bright enough in darkness, or be otherwise hard to read. If your phone set doesn't have a display that you like, you should try to exchange it. Ideally, you should check out all the different displays when you're in the "cell phone store" shopping for a set.
- In the physical sense, some cell phones are a great deal more rugged than others. You can drop some sets onto a concrete floor from shoulder height and they'll suffer no damage. But few sets can take that kind of abuse, and some are so fragile that you might wonder whether they're meant for use in weightless environments only.

Quick Question, Quick Answer

- What's the PIN lock feature on a cell phone? Who needs to use it?
- The PIN (personal identification number) lock setting makes it practically impossible for anyone to use your cell phone in case you lose it. You should activate the PIN lock if you want to keep your cell phone account, and any personal information that you might have stored on the phone, secure. Experts recommend that everybody who has a cell phone "lock it down" with a PIN code right away after buying it!

Fact Or Myth?

- Does extensive use of a cell phone increase the risk of brain cancer?
- Experts disagree on this issue. Some studies seem to show a correlation between cell phone use and brain tumors. However, other studies show no correlation, and a few have come up with a negative correlation, as if cell phone use might actually protect the brain! I don't know the answer to this mystery.

Warning!   Cell phone use is illegal in certain locations and situations. Always make sure that you know when you're in one of those scenarios, and switch that little thing off. Also, in some parts of the world (and increasingly in the United States), you can't legally use a cell phone while driving a motor vehicle.

Wireless Tablets

In recent years, *tablet devices* have grown immensely popular. The simplest ones, called *e-book readers*, are meant only to display *electronic books* (e-books) downloaded from the Internet with a computer and then transferred to the tablet with a special cable. Some of these devices can access the Internet by means of *Wi-Fi*, which employs wireless routers located in homes and businesses with Internet access. Examples of Wi-Fi-equipped e-book readers include Amazon's Kindle Fire and Barnes and Noble's Nook Tablet (as of late 2012). The most sophisticated tablet devices comprise small computers with touch screens having diagonal measures of 7 to 12 inches. The flagship example is the Apple iPad. Similar devices are manufactured by Microsoft, Samsung, and others. They can access the Internet in either or both of two ways: by means of Wi-Fi or by means of the cellular network.

How Wi-Fi Works

The term Wi-Fi comes from the expression "wireless hi-fi." Since its origins, Wi-Fi has evolved into a technology used almost entirely with personal computers and tablets. Basically, if you connect a wireless router to a regular Internet modem, such as the one you get as part of a cable or satellite installation, you have your own *Wi-Fi hotspot*. The maximum working range of a wireless router rarely exceeds 100 feet (about 30 meters), and unless you have an ideal installation, you shouldn't expect it to cover a radius of more than half that.

These days, you'll find Wi-Fi hotspots in public libraries, hotels, motels, restaurants, bars, and airports. Even a few fast-food places and department stores have them. You can bring your notebook computer or Wi-Fi-equipped tablet device into such a place, obtain the password from one of the employees, and get on the Internet, sometimes with quite respectable download and upload speeds. The quality of a Wi-Fi hotspot connection depends on the quality of the router that the establishment has installed, the speed of their own Internet connection, and the number of other users taking advantage of the hotspot along with you.

When you use any public Wi-Fi hotspot, you should keep in mind the fact that your device, whether it's a sophisticated notebook computer or a simple e-book reader or anything in between, is effectively networked (connected to) all the other devices using the same hotspot. If one of those other users is somebody who likes to snoop around in other people's Internet business and has the technical knowledge to do it (or, to use the standard jargon, a *hacker*), then he or she can not only observe everything you do online, but also take over your computer or corrupt its contents.

Did You Know?

Wi-Fi hacking doesn't happen very often (in terms of the proportion of hacked user-hours to total user-hours), but in major hotels serving a worldwide clientele, or in international airports, you should not be surprised if your computer starts to "run itself" while you're online in the hotspot. If that happens, switch your computer or tablet device off immediately, even if you have to force the process by holding down the power button. Paradoxically, you should consider yourself lucky if you notice a hack. In most cases, unless you're an experienced techie, you'll get no clue about the intrusion until later (when your device doesn't work properly anymore). If the hacker wanted only to use your device as a "robot" or "zombie" to send junk e-mails or conduct other illicit business through your machine, you might never find out about the hack at all.

Woes and Resolutions

Aside from the potential problems you might have with hackers at major hotspots, you might encounter one or more of the following glitches.

- In densely populated areas, some people drive or walk around with portable computers, looking for unsecured wireless routers at residences (hopefully never yours) so they can access the Internet through them. Use a good password with your wireless router. A good password comprises letters and numerals in a random sequence. Never operate any wireless router without a password. Although passwords can't guarantee protection against roving hackers, they greatly minimize the risk because these nosy folks usually look for easy (unsecured) targets and skip over the more difficult (secured) ones.

- Once in awhile, you'll find that a device won't access the wireless router when you have another device online with it already. I have two Wi-Fi-equipped e-book readers. When one of them is online, the other one will usually connect, but not always! The only cure for this trouble, should you ever experience it, is to avoid using both of the incompatible devices with Wi-Fi at the same time.

- If your router is located in a bad spot, surrounded by too many obstructions or other electronics devices, you might get far less than the advertised working range for your home hotspot. Try relocating the router to a room near the center of the house or on an upper floor. Don't put it in the basement, though.

- In certain locations, other wireless devices' signals will create so much interference that your router's signal is drowned out except when you use a tablet or computer within a few feet of it. If you can find the source of the interference and power it down, consider yourself lucky. In most cases you'll never manage to isolate or identify the source, or even if you do, you won't be able to do anything about it.

- Once in awhile, a router will lose contact with the modem that serves it, and won't be able to "figure out" how to identify it and connect with it again. In that case, switch off the router and the modem, wait a couple of minutes, then switch the modem back on until it connects to the Internet, and finally switch the router back on until it connects to the modem. You'll be able to tell when a modem or router has established its connection by watching the lights on it. A certain pattern of steady and flashing lights will tell you that the device has connected. The exact pattern varies from device to device; you'll get familiar with yours after you've used it for awhile.

For Nerds Only

If you have a portable AM/FM radio receiver and you can hear a buzz or whine on it when you tune it to a vacant channel, then you know that you have an *electromagnetic (EM) noise source* in your vicinity. An awful lot of things can generate EM noise, which can adversely affect sensitive wireless systems including Wi-Fi arrangements. If you're plagued with EM noise, you might be able to locate its origin with the help of a portable radio that "hears" it. Rove until you come up to the point where the noise reaches overwhelming strength. At that time, the source should reveal itself. Whether or not you can turn off the offending device, however, will depend on what it is. You can turn off a wireless "baby monitor" a lot more easily than you can shut down a transformer on top of a utility pole!

The Global Positioning System

The *Global Positioning System* (GPS) is a network of radiolocation and radionavigation apparatus that operates on a worldwide basis. The system employs numerous satellites, and allows you to determine your location on the earth's surface, and in some cases, your altitude above the surface as well.

How It Works

All GPS satellites transmit signals that have wavelengths on the order of a few inches. The signals carry special codes that contain timing information used by the receiving apparatus to make measurements. A GPS receiver determines its location by measuring the distances to three or four satellites (preferably four) by precisely timing the signals as they travel between the satellites and the receiver. The process resembles the *triangulation* used in old-fashioned navigation, except that with the GPS, it takes place in three-dimensional space rather than on the two-dimensional surface of the earth.

The GPS receiver uses a computer to process the information received from the satellites, which follow circular orbits several thousand miles up. From this information, the GPS unit can give you an indication of your geographical position. For individual users, the accuracy of the positioning readout varies slightly, depending on the relative positions of the satellites with respect to the user's location. The larger the number of satellites involved, the more accurate the readout.

Because it's a radio receiver, a GPS unit always requires an antenna. The type of antenna depends on the situation. Figure 7-6 shows two common antenna types found on cars and trucks. The whip design, shown at A, can have a magnetic mount similar to the hardware used with CB and small amateur radio antennas, or it can have a mounting that sticks to a rear window (if that window is near enough to horizontal). The streamlined design, shown at B, is more damage-resistant than the larger type, but doesn't offer as much sensitivity. More sophisticated antennas are needed for boats and aircraft; they must withstand corrosive environments or extreme airspeeds. Handheld GPS units have antennas built-in, and look like cell phone sets.

An increasing number of new automobiles, trucks, and boats have GPS receivers preinstalled. If you are driving in a remote area and you get lost, you can use the GPS system to locate your position. Then, with the aid of a cellular telephone (or an amateur or CB radio, if you're out of cell phone range), you can call for help and inform authorities of your location. Arguably, every motor vehicle and boat should have preinstalled GPS and cellular communications equipment to enhance the safety of passengers.

For Nerds Only

Radio signals travel at 186,282 miles per second (299,792 kilometers per second) through space and the atmosphere. A slight reduction in wave-propagation speed occurs in the *ionosphere* (upper atmosphere with ionized atoms). The extent of this speed reduction depends on the signal frequency. The GPS employs two-frequency transmission to compensate for this effect. The delay difference is fed to a computer, which generates a correction factor to cancel out the error.

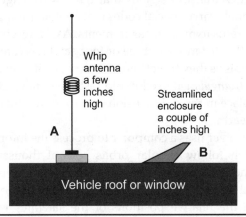

FIGURE 7-6 Automotive GPS antennas. At A, whip type; at B, low-profile type.

Woes You Can Resolve by Yourself

Unless you're a professional electronics technician, you won't be able to do anything "on your own" about malfunctioning GPS equipment in your vehicle, unless the trouble is caused by a radio transmitter. If that's the case, you'll know right away when you press the microphone button and start talking!

Your GPS system shouldn't have any problems with the signals from CB radios because your CB transmitter outputs only a few watts unless you're a *freebander* (one of those folks who flouts the law and uses an amplifier).

If you're one of the few amateur radio operators who have a high-power transmitter in a motor vehicle or on a boat, you might experience GPS glitches when you transmit on certain frequencies. The only sure-fire cure for this problem is to avoid transmitting with the radio when you want to use your GPS system!

Woes You Can Resolve with Help (Hopefully)

Global Positioning System users complain fairly often about three problems aside from conflicts with mobile radio transmitters. You can sometimes resolve these bugaboos with the help of service technicians. In other cases, you'll have to get rid of your old system and buy a new one.

1. *Incorrect coordinates or location display*—Unless the receiver can clearly "see" at least three (and preferably four) satellites, your system might show you an incorrect location. In most cases the discrepancy will be so great as to make itself obvious. For example, you might be driving on a Texas highway, but your system will tell you that you're in Saskatchewan. A good system will give no reading unless it can get good signals from the necessary number of satellites. Once in awhile, a bad antenna or a defective receiving unit will cause you to see an improbable or ridiculous location display or set of coordinates. A technician can check your receiver and antenna to see if they're working okay, and can suggest repairs or modifications in case they aren't.

2. *Slow responsiveness*—If your GPS unit is "competing" with a lot others in your general area, you'll observe a long delay in the response, or your system will tell you where you were a few minutes ago, rather than where you are at the moment. There's not much you can do in a situation like this if you're in motion, except pull over to the side of the road and wait until your GPS display "catches up." You can be reasonably sure that it's "caught up" to you when it, like you, comes to rest! This problem is most likely to occur in and around large cities where GPS and other wireless usage is heavy.

3. *"Frozen" display or coordinates*—If your GPS receiver loses contact with the network, the coordinates or display will "freeze" or even disappear altogether. If you're driving along a highway at high speed and you're used to watching a moving position indicator, and instead you see that it has

come to a stop, you'll know that something has gone wrong. The receiver might have moved out of a line of sight with enough satellites, or you might have a weak receiver or a bad antenna. If you experience the problem often, or when you wouldn't expect a loss of signal, or if it happens more and more often in the same general location, you should take your system to a technician and have it checked out.

Access Control

Wireless technology is used in access-control devices, circuits, and networks. Systems range from simple machines, such as card readers or keypad-entry devices, to sophisticated electromechanical networks. Let's look at three examples of wireless devices that can help you protect your property from unwanted human visitors.

Knowledge-Based Access Control

In a *knowledge-based security system*, authorized people are issued numerical codes. The entrances to your property have locks that disengage when the proper sequence of numbers is punched into a keypad. This keypad can be hard-wired into the system, or it can be housed in a box about the size of a cell phone. It works like a bank automatic-teller machine (ATM) personal identification code.

You might decide that you need only one access code, which you can give to all the people that you want to authorize. Alternatively, you can issue a different code to each authorized person. The term "knowledge-based" arises from the fact that, in order to gain entry to your property, a person must know a specific piece of information (in this case the access code).

One of the main advantages of this type of system is that the codes cannot easily be guessed. The authorized people should memorize their access numbers. The numbers should never be written down in any form that will give away their meaning or purpose. Another asset of knowledge-based security systems is their relatively low cost.

A disadvantage of this scheme is that access codes occasionally leak out. People tend to give secrets away when situations arise that make it expedient to do so (or if it becomes inexpedient not to). Also, codes are sometimes forcibly stolen. Once a code is stolen, anyone who has it can get into the property until that code is invalidated.

Possession-Based Access Control

A *possession-based security system* requires authorized people to possess some physical object that unlocks the entry to your property. Magnetic cards are a popular form of possession-based security device. You insert the card into a slot, and a microcomputer reads data encoded on a magnetic strip. This data can be as simple as an access code of the sort you punch on a keypad. Or it might contain many details about you. A so-called *smart card* can be used for security purposes. So can bank ATM cards, credit cards, and radio-frequency identification (RFID) cards.

A *passive transponder* provides a wireless form of possession-based security system. It's a magnetic tag that authorized people can wear or carry. They're the same little things that department stores employ to deter petty thieves. The transponder can be read from several feet away.

A *bar code tag* is another form of passive transponder. You'll see them in stores, where they are used for pricing merchandise and keeping track of inventory. Bar coding allows instant *optoelectronic identification* of objects. A bar-code tag has parallel bands of various widths. More sophisticated tags have complex patterns of black shapes on a white background. A laser rapidly scans the pattern. The dark regions absorb the laser light, while the white regions reflect the light back to a sensor. The sensor thereby receives a binary data signal unique to the pattern on the tag. This signal can contain considerable information even when the pattern on the tag seems simple.

The main advantage of possession-based security systems is convenience. You need not worry about forgetting a code number. But this advantage is more or less nullified by the fact that little plastic cards can easily get misplaced, and they're easy to steal. They're also easy for unscrupulous store employees to "catch and copy."

Did You Know?

If you've ever been leaving a store or library and been detained because an electronic detector beeped at you, you've experienced a passive transponder at work!

Biometric Access Control

A *biometric security system* gets its name from the fact that it detects and acts on certain biological characteristics of people authorized to enter a property. For example, it might employ a camera along with a pattern-recognition computer program to check a person's facial contours against information in a gigantic database. The machine might use speech recognition to identify people by breaking down the waveforms of their voices. It might record a hand print, fingerprint, or iris print. It might even employ a combination of all these things. A computer analyzes the data obtained by the sensors, and determines whether or not the person is authorized to enter the premises.

Wherever security requirements are so strict that a biometric security system is deemed necessary, a few brilliant, reckless rogues intent upon defeating it and getting into the premises will doubtless exist, waiting patiently, continuously, and endlessly for an opportunity to strike. A government installation in a hostile country is a prime example of such a property. While the system can be almost impossible to fool, it must also be set up so that a brute-force surprise attack will not likely overcome it. History has shown that this level of security is nigh impossible to attain, and at the same time, ensure that no authorized personnel are falsely arrested or injured as a result of a system error.

For homeowners and small businesses, biometric systems are generally too expensive. But exceptions to this rule do occur. Top-secret archives, priceless works of art, and some scientific experiments (and research personnel) justify the most sophisticated security systems available, no matter what the cost.

Intrusion and Fire Detection

In addition to access control, you might want to install a security system that can detect the presence of an intruder who happens to "get past the gate." Such systems can interconnect with telecommunications networks to alert the police when a security breach occurs, or the fire department in case of a fire.

Electric Eye

The simplest device for detecting an unwanted visitor (besides a smart dog) is an *electric eye*. Narrow beams of visible light shine across all reasonable points of entry, such as doorways and window openings. A photodetector receives energy from each beam. If the photodetector stops receiving its assigned beam, an alarm is actuated. The main problem with this system is that a power failure can trigger the alarm unless the entire system has a backup battery.

The typical electric eye has a light source, usually a laser diode, and a light sensor, such as a photoelectric or photovoltaic ("solar") cell. These devices are connected into an actuating circuit, as shown in Fig. 7-7. When something interrupts the light beam, the voltage or current passing through, or generated by, the sensor changes dramatically. An electronic circuit detects this voltage or current change. Using amplifiers and switches, even the smallest change can be harnessed to control massive machines.

For Nerds Only

Electric eyes often use infrared (IR), rather than visible light. This methodology works especially well in burglar alarms because an intruder can't see an IR beam, and therefore can't evade it. However, an IR-based electric eye might get fooled by the presence of a hot object, such as a lighted match or cigarette, in the line of sight of the receptor.

Infrared Motion Detector

Many popular intrusion alarm devices make use of infrared (IR) motion-detection transducers. Two or three wide-angle IR pulses are transmitted at regular intervals; these pulses cover most of the room in which the device is installed. A receiving

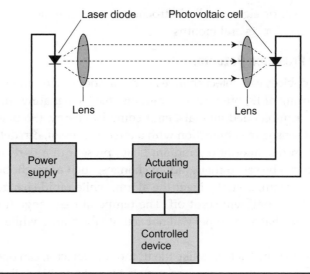

FIGURE 7-7   Functional diagram of an electric eye.

transducer picks up the returned IR energy, normally reflected from the walls, the floor, the ceiling, and the furniture, as shown in Fig. 7-8. The intensity of the received pulses is measured and recorded by a microprocessor. If anything in the room changes position, the intensity of the received energy will vary, and the resulting signal will set off an alarm. These devices consume very little power in regular operation, so batteries can serve as the power source. A typical alarm system of this

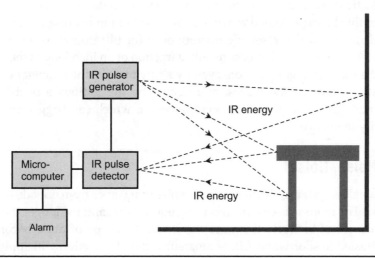

FIGURE 7-8   Functional diagram of an infrared (IR) motion detector.

kind uses six or eight small electrochemical cells, which will operate the device continuously for several months.

Infrared Presence Detector

Infrared devices can detect changes in the indoor environment in another way: the direct sensing of IR, often called "radiant heat," emanating from objects. Humans, and all warm-blooded animals, emit some IR energy. (So does fire, of course.) A simple IR sensor, in conjunction with a microprocessor, can detect rapid or sudden increases in the amount of "radiant heat" present in a particular place. The time threshold can be set so that gradual changes, such as might be caused by the sun warming a room, do not trigger the alarm, while rapid changes, such as a person entering the room, will set it off. The temperature-change (increment) threshold can be set so that a small pet will not actuate the alarm, while a full-grown person will do it.

An IR presence detector, like the IR motion detector, can operate from batteries, and in fact, consumes even less power than the motion detector because no IR pulse generator is required. The main problem with "radiant heat" detectors is that they can be fooled. False alarms pose a significant risk; for example, the sun might *suddenly* come out on a cloudy day and shine directly on the sensor and trigger the alarm. It's also possible that a person clad in a winter parka, boots, hood, and face mask, entering from a subzero outdoor environment, might fail to set off the alarm. For this reason, radiant-heat sensors are used more often as fire detectors and alarm actuators than as intrusion detectors.

Ultrasonic Motion Detector

Motion in a room can be detected by sensing the changes in the relative phase of acoustic waves. An ultrasonic motion detector contains a set of *transducers*, which resemble loudspeakers that work above the human hearing range. Another set of transducers, which resemble microphones for ultrasound, picks up the reflected acoustic waves, which measure only a fraction of an inch long in the air. If anything in the room changes position, even by a tiny bit (like the diameter of a penny), the relative phase of the waves, as received by the various acoustic pickups, will change. This data goes to a microprocessor, which can trigger an alarm and/or notify the police.

Noise, Noise, Noise!

In wireless systems, electromagnetic noise that comes from outside is called *external noise*. The more sensitive the receiving equipment, and the longer the distance over which it has to work, the more significant this type of noise becomes. If you're interested in shortwave, CB, or amateur radio, this section will apply especially to you! But as ordinary household wireless devices grow more sophisticated and sensitive, the following terms and concepts will become important to everybody.

Put your techie hat on, and let's forge ahead into the deep realms of Brother Edsel Murphy, the (fictitious) mad scientist who coined *Murphy's Law*.

Murphy's Law

The short, and most commonly invoked, version of Murphy's Law states that "If something can go wrong, it will." Some of the more cynical technical people among us might add a corollary that says "If something *cannot* go wrong, it will."

Cosmic Noise

Noise from outer space is known as *cosmic noise*. It occurs all throughout the entire electromagnetic spectrum, from the very-low-frequency (VLF) radio band where waves are kilometers long to the realm of x rays and gamma rays where the waves measure a tiny fraction of a millimeter long. At the lower frequencies, the ionized upper atmosphere of our planet prevents the noise from reaching the surface. At some higher frequencies, the lower atmosphere prevents the noise from reaching us. But at many frequencies, cosmic noise arrives at the surface at full strength.

Cosmic noise can be identified by the fact that it correlates with the plane of the *Milky Way*, our galaxy. The strongest *galactic noise* comes from the direction of the constellation *Sagittarius* ("The Archer") because this part of the sky lies on a line between our Solar System and the center of the galaxy. Galactic noise was first noticed and identified by *Karl Jansky*, a physicist working for the Bell Laboratories in the 1930s. Jansky conducted experiments to investigate and quantify the earth's atmospheric noise at a wavelength of about 15 meters, or a frequency of 20 MHz. He found some radio noise that he couldn't account for, and then he noticed that its orientation correlated with the location of the Milky Way in the sky. Jansky's antenna was a simple affair like the ones used by amateur radio operators.

Galactic noise, along with noise from the sun, the planet Jupiter, and a few other celestial objects, contributes to most of the cosmic noise arriving at the surface of the earth. Other galaxies radiate noise, but because those external galaxies lie much farther away from us than the center of our own galaxy does, sophisticated equipment is needed to detect the noise from them.

Did You Know?

Radio astronomers deliberately listen to cosmic noise in an effort to improve their understanding of the universe. To them, humanmade noise and interference (including legitimate broadcast and communications signals), rather than cosmic noise, constitute the major nuisances.

> **Here's a Tale!**
>
> In 1965, *Arno Penzias* and *Robert Wilson* of the Bell Laboratories observed cosmic noise that seemed to come from everywhere. For some time, the noise source remained a mystery. Nowadays, most astronomers believe that the noise originated with the fiery birth of our universe (an event often called the *Big Bang*), and comes to us "delayed" by billions of years! If they're right, then when we detect and record this noise, we in effect "hear" the echo of Creation.

Solar Noise

The amount of radio noise emitted by the sun is called the *solar radio-noise flux*, or simply the *solar flux*. The solar flux varies with frequency. But no matter what the frequency (or wavelength), the level of solar flux increases when an eruption on the sun's surface, known as a *solar flare*, occurs. A sudden increase in the solar flux indicates that "shortwave radio" broadcasting and communications conditions (including amateur radio) will deteriorate within a few hours.

The solar flux is commonly monitored at a wavelength of 10.7 centimeters, which corresponds to a frequency of 2800 MHz. At this frequency, which is about 100 times the frequency of everyday CB radios, the earth's atmosphere has little or no effect on radio waves, so the energy reaches the surface at full strength.

> **For Nerds Only**
>
> The 2800-MHz solar flux is correlated with the 11-year *sunspot cycle*. On the average, the solar flux is highest near the peak of the sunspot cycle, and lowest near a sunspot minimum. As of this writing, you can read data about solar activity by visiting the website of the American Radio Relay League at www.arrl.org. Look for a link dealing with the "solar update."

> **Fact or Myth?**
> - In recent years, we've heard about solar flares and the potential danger that their effects pose to electronic devices. Some people say that these events occur more often during sunspot peaks than at other times. Are they telling the truth?
> - Yes, but solar flares can occur at any time.

Atmospheric Noise

Electromagnetic noise is generated in the atmosphere of our planet, mostly by lightning discharges in thundershowers. This noise is called *sferics*. In a radio receiver, sferics cause a faint background hiss or roar, punctuated by bursts of sound we call "static." Figure 7-9 shows an example of sferics as they might look on the display of a laboratory oscilloscope connected into an AM radio receiver.

A gigantic voltage constantly exists between the surface of the earth and the ionosphere. The earth's surface and the ionosphere behave like concentric, spherical surfaces of a massive capacitor, with the troposphere and stratosphere serving as the insulating material (called a *dielectric*) that keeps the charges separated. Sometimes this dielectric develops "holes," or pockets of imperfection, where discharge takes place. Such "holes" are usually associated with thundershowers. There are normally about 700 to 800 such areas on the earth at any given time, concentrated mostly in the tropics. Sand storms, dust storms, and volcanic eruptions also produce some lightning, contributing to the overall sferics level.

Sferics are not confined to our planet! A great deal of radio noise is generated by storms in the atmosphere of the planet Jupiter. Astronomers can "hear" this noise with radio telescopes. Sferics probably also occur on Saturn, and perhaps on Uranus, Neptune, Venus, and Mars as well. In the cases of Venus and Mars, dust storms and volcanic eruptions would likely be the cause of sferics.

For Nerds Only

You can hear the sferics from a distant thundershower on the standard AM broadcast band. If you have a general-coverage "shortwave" communications receiver, you can listen at progressively higher frequencies as the storm or storm system approaches. When you hear the static bursts at 30 MHz, the storm is probably less than 100 miles away from your location.

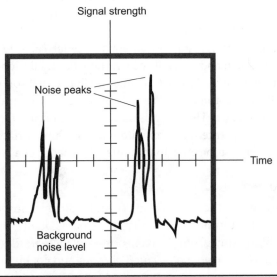

FIGURE 7-9   If you connect an oscilloscope into an AM radio receiver and listen to sferics, you'll see a display that looks something like this.

Precipitation Noise

Precipitation noise, also called *precipitation static*, is radio interference caused by electrically charged water droplets or ice crystals as they strike metallic objects, especially antennas. The resulting discharge produces wideband noise that sounds similar to the noise generated by electric motors, fluorescent lights, or other appliances.

Precipitation static is often observed in aircraft flying through clouds containing rain, snow, or sleet. But occasionally, precipitation static occurs in radio communications installations. This is especially likely to happen during snow showers or storms; then the noise is called *snow static*. Precipitation static can sometimes make radio reception difficult or nearly impossible, especially at low frequencies (long wavelengths).

In a good radio receiver, a *noise blanker* or *noise limiter* can reduce the interference caused by precipitation static. A means of facilitating electrical discharge from an antenna, such as an inductor between the antenna and ground, can also help. If the antenna elements have sharp points on the ends, you can blunt them by installing small metal spheres on those ends, although if you do that, you might have to shorten the elements slightly in order to account for *loading effects* that the spheres produce.

Corona

When the voltage on an electrical conductor (such as an antenna or high-voltage transmission line) exceeds a certain level, the air around the conductor begins to ionize. That means the atoms in the air gain or lose electrons, so that they become electrically charged. If the effect becomes significant, it can cause a blue or purple glow called a *corona* that can be seen at night. This glow commonly occurs at the ends of a radio broadcast or communications transmitting-antenna element at night when the transmitter has high power output. Coronae occur increasingly often as the relative humidity rises because it takes less voltage to ionize moist air than it takes to ionize dry air. A corona produces a strong, sometimes overwhelming hiss or roar in radio receivers.

A corona can occur inside an antenna's feeder cable just before the dielectric material breaks down and the cable gets ruined. Poorly designed antennas with high-power transmitters can subject a transmission line to that sort of stress. So can nearby thundershowers. A corona is sometimes observed between the plates of capacitors handling large voltages. This effect is more likely to occur with a pointed object, such as the end of a whip antenna, than with a flat or blunt surface. Some antennas have small metal spheres at the ends to minimize the chance of a corona occurring.

Did You Know?

Coronae sometimes occur as a result of high voltages caused by static electricity during thunderstorms. Such a display is occasionally seen at the tip of the mast of a sailing ship. Coronae were observed by seafaring explorers hundreds of years ago. They called them *Saint Elmo's fire*, and not knowing about electricity, they attributed them to supernatural powers.

Impulse Noise

Any sudden, high-amplitude voltage pulse will generate an electromagnetic field, and radio receivers will often pick it up. It's called *impulse noise,* and it can come from all kinds of household appliances, such as vacuum cleaners, hair dryers, electric blankets, thermostats, and fluorescent-light starters. Impulse noise tends to get worse as the frequency goes down, and can plague AM broadcast receivers to the consternation of their users. Serious interference can occur in "shortwave" radio receivers, but it gets less severe as the frequency rises, and it rarely poses a problem above 30 MHz.

Impulse noise in a radio receiver can be reduced by the use of a good ground system. All the components in the system should be grounded by individual wires to a single point. A noise blanker or noise limiter can also help, if the receiver has one. A "shortwave" or amateur radio receiver should be set for the narrowest response bandwidth consistent with the mode of reception.

Impulse noise can be picked up by high-fidelity (hi-fi) audio systems. The greater the number of external peripherals (such as computers, flash-memory chips, microphones, and speakers) that exist in a hi-fi system, the more susceptible the system becomes to this type of noise. Wireless microphones and wireless headsets are especially vulnerable to its whims. An excellent ground connection is imperative. You might have to shield all speaker leads and other interconnecting conductors, using coaxial cable instead of "open wire."

Ignition Noise

Ignition noise is impulse noise generated by the electric arcs in the spark plugs of an internal combustion engine. Many different kinds of devices produce it, including automobiles and trucks, lawn mowers, and gasoline-engine-driven generators. Figure 7-10 shows how ignition noise looks on an oscilloscope connected into the sensitive amplifiers of a radio receiver.

In a "shortwave" or amateur radio receiver, a noise blanker can often work wonders to get rid of ignition noise problems. The pulses of ignition noise are of very short duration, although their peak intensity can be considerable. Noise blankers are designed to literally switch the receiver off during these short noise pulses.

As you learned earlier in this book, ignition noise often poses a problem for mobile two-way radio operators, especially if communication in the "shortwave" band is contemplated. Ignition noise can be worsened by radiation from the distributor wiring in a truck or automobile. Sometimes, special spark plugs, called *resistance plugs,* will greatly reduce the amount of ignition noise that an engine produces. A competent automotive technician should know whether or not this option will work for you. An excellent vehicle-chassis ground connection is imperative in any mobile installation.

Ignition noise is not the only source of trouble for the mobile radio operator. Noise can also be generated by the friction of the tires against the pavement, especially in dry, hot weather. High-tension power lines often radiate significant

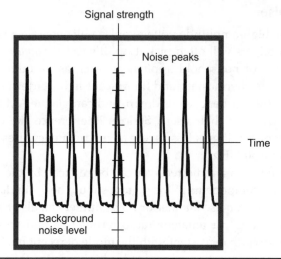

FIGURE 7-10   Here's what ignition or impulse noise looks like on an oscilloscope display connected into an AM radio receiver.

impulse noise. The vehicle's alternator can cause impulse noise in the form of a whine that changes pitch as the car accelerates and decelerates.

Power-Line Noise

Utility lines, in addition to carrying the 60-Hz AC that they're meant to transmit, carry other currents. These currents have a broadband nature. They result in an effect called *power-line noise*, which is a particularly powerful form of dirty electricity. The "rogue currents" usually occur because of electric *arcing* at some point in the utility grid. The arcing might originate in household appliances; it can take place in faulty or obstructed utility transformers; it can occur in high-tension lines as a coronal discharge into humid air. The currents cause the power line to radiate EM fields like huge radio transmitting antennas!

Power-line noise sounds like a buzz, hiss, or roar when picked up by a radio receiver. Some types of power-line noise can be attenuated by means of a noise blanker. Other types of dirty electric noise defy noise blankers, and the best you can do is hope that an *automatic noise limiter* (ANL) will give some relief by at least giving the desired signals a "fighting chance" against the noise. Engineers have gone to great lengths, racking their brains to come up with new and innovative ways to deal with dirty electric noise as it becomes more and more of a problem.

For Nerds Only

In a scheme called *phase cancellation*, noise from a special auxiliary antenna is used to null out the noise from the main receiving antenna, in effect making the noise destroy itself. The auxiliary antenna actually works best if it *does not* pick up any of the desired signals! I've used that technique at very low frequencies, and it can sometimes work amazingly well when everything else fails. The trick lies in making sure that the "noise antenna" picks up lots of noise but little or no signal, and that its output can be adjusted so that it exactly "bucks" the noise picked up by the main antenna.

Did You Know?

When you ponder all the wireless devices that transmit and receive EM signals in a typical household or business, you'll realize that it would be a miracle if conflicts did *not* occur from time to time! Waves upon waves upon waves (Fig. 7-11) permeate the space all around us, and that's a prescription for trouble. As technology advances and wireless devices continue to proliferate almost out of control, we can only expect that this scenario will grow uglier and uglier.

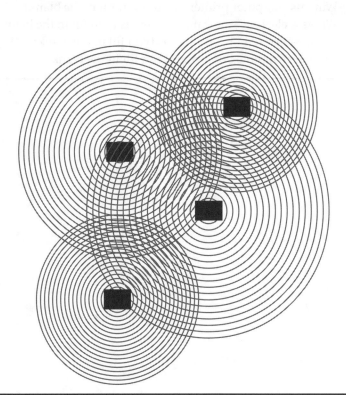

FIGURE 7-11   Electromagnetic waves from multiple wireless devices flow around and through us all the time, wherever we go.

Try These Experiments!

Here's an "electronic odyssey" that will keep you busy for a while. Do a bunch of "tweak freak tests" with wireless devices, two-by-two, and see what interferes with what. You can start with these ideas:

- Wireless headset versus (you fill in the blank)
- Desktop computer versus (you fill in the blank)
- Cell phone set versus (you fill in the blank)
- Wireless tablet computer versus (you fill in the blank)
- Cordless phone set versus (you fill in the blank)
- Compact fluorescent lamp versus (you fill in the blank)
- Light-emitting-diode (LED) lamp versus (you fill in the blank)
- Light dimmer versus (you fill in the blank)
- Television set versus (you fill in the blank)
- FM radio versus (you fill in the blank)
- AM radio versus (you fill in the blank)
- Wireless computer mouse or keyboard versus (you fill in the blank)
- Wireless computer printer versus (you fill in the blank)
- Wireless electric "smart meter" versus (you fill in the blank)
- Garage-door remote box versus (you fill in the blank)
- Appliance remote box versus (you fill in the blank)

CHAPTER 8

What Else Can You Do?

If you want to fill your life and home with gadgets and gimmicks, electronics offers opportunities galore. In this chapter, you'll learn about some common ways that electronics can improve and enrich your life. You'll also get a chance to build a simple wet cell and then do a couple of experiments on your own body!

Try CFLs

Fluorescent lamps offer better efficiency than incandescent light bulbs do, but you can't remove an incandescent bulb from its socket and put a conventional fluorescent "tube" in its place. Engineers invented the *compact fluorescent lamp* (CFL) to satisfy people's desire for a direct-replacement fluorescent counterpart to the incandescent bulb.

How They Evolved

Fluorescent and incandescent lamps differ in several ways. One of the most obvious, but also most easily overlooked, differences lies in the amount of light that the device produces per unit surface area. An incandescent lamp has a brilliant filament with a tiny surface area that emits all of the light, but the light from a fluorescent lamp comes off uniformly from a surface with considerable area. This difference presented the inventors of the CFL with a major challenge: How could they make a fluorescent light bulb physically small without sacrificing overall brilliance?

A typical "old-fashioned" fluorescent tube has the shape of a long, thin cylinder. If the entire assembly is coiled up into a helix, the overall size of the lamp can be reduced. However, the amount of light emitted per square millimeter of bulb surface must increase if the lamp is to maintain its overall brilliance. The *phosphor* (coating on the inside of the glass that makes the whole thing glow) in a CFL must, therefore, produce far more visible light per unit of area than the phosphor in a conventional fluorescent bulb does. As a result, CFLs burn hotter than conventional fluorescent bulbs do. The CFL phosphor must withstand the high temperatures without rapidly degenerating.

The CFL concept originally came about during the "energy crisis" of the 1970s. Engineers at General Electric produced a working CFL but decided against mass production because of the high startup cost. For consumer use, CFLs didn't gain traction until the mid-1990s when China began exporting helical fluorescent tubes, which gave off unnatural blue-green light. By around 2005, this problem had been overcome. Today, you can get CFLs that produce light almost indistinguishable from the output of a typical incandescent bulb.

CFLs versus Incandescent Lamps

If you see two otherwise identical table lamps with shades, one having a 60-watt "soft-white" incandescent bulb and the other having a CFL that produces the same amount of light output, you probably won't know which bulb is which until you look inside the shade. But a "60-watt-equivalent" CFL consumes only 10 to 12 watts of actual power. You can expect the CFL to last several times longer than an incandescent bulb does, although the CFL costs more up front. Savings accrue over time as the result of less frequent bulb replacement and reduced overall energy consumption. In the end, a good CFL gives you the better deal.

One of the most significant problems with CFLs arises from the fact that they don't work very well in cold weather. If you live in a region where the winter temperatures drop below freezing, you won't like CFLs for outdoor use. They'll have trouble starting up, and once they do get going, they'll take awhile to reach full brightness. If the temperature gets much below 0°F (−18°C), as it can in the northern United States and most of Canada, CFLs might not start up at all. If you want to replace outdoor incandescent lamps with more efficient devices, you'll do better to go with *light-emitting-diode* (LED) lamps.

Another, less noticeable problem with conventional CFLs is a gradual decline in light output as they age. The newer designs suffer less from this trouble than the earliest ones did; some people don't even notice it until the old CFL burns out and they install a new one of the same wattage. While you can rely on CFLs to last longer than incandescent lamps do, the lifespan difference is greatest in situations where you don't turn the lamp on and off very often. Yet another limitation of CFLs is the fact that the basic types won't work with dimmers. You can buy dimmable CFLs, but they cost more than ordinary ones.

Did You Know?

You must use care when disposing of a burned-out CFL because it contains a small amount of mercury. Rules for disposal vary depending on where you live. If you break a CFL, you should avoid direct exposure to the contents when cleaning up the mess. You don't have to wear a hazmat suit, but you should don a dust mask to make sure that you don't inhale any of the airborne dust and debris. Those masks don't cost much; you can buy disposable ones at almost any decent department store or pharmacy. You should also wear disposable gloves (also cheap, and also available at department stores or pharmacies) and wash your hands after you've finished the cleanup task. Follow the instructions provided by the bulb manufacturer to prevent leakage of mercury into the environment, or call your local garbage removal service provider and ask them what to do with burnt-out CFLs. Demised-CFL disposal precautions are much the same as they are for automotive batteries that contain lead.

Radio-Frequency Interference

In some situations, CFLs produce high-frequency AC that can interfere with wireless-device reception at close range. Do you have a radio-controlled garage-door opener whose motor box has a light that comes on for a few minutes after you close the door? Is it an incandescent bulb? If so, try replacing that incandescent lamp with a CFL, and see if you still get the same operating range that you did before.

I discovered the *radio-frequency interference* (RFI) from a CFL by accident. The remote-control box for my garage-door opener transmits a radio signal to the motor. It normally works up to 100 feet away. When I replaced the 60-watt incandescent lamp in the motor box with a "60-watt-equivalent" CFL, the range went down to 20 feet. Because I'm an amateur radio operator, and therefore, have had plenty of experience with RFI issues, I diagnosed the trouble right away. I put a new 60-watt incandescent lamp in the motor box, and the system worked normally again.

For Nerds Only

If you notice performance issues with wireless devices (cordless phones, wireless routers, and the like) after replacing a lot of the incandescent lamps in your house with CFLs, try putting incandescent lamps back into the fixtures one by one, and see if the problem goes away.

Long Question, Long Answer

- Let's say that your regional utility company charges 10 cents per kilowatt hour (kWh) of electricity usage. If you replace two dozen 60-watt incandescent lamps with "60-watt-equivalent" CFLs that consume only 12 watts each, and if you burn the lamps for an average of 6 hours per day each, how much money can you save over the course of a year, ignoring the up-front cost of the CFLs and the replacement costs of burned-out incandescent lamps? (Most incandescent lamps would likely burn out within a year, but only a small fraction of the CFLs will.)

- Keep in mind the difference between energy and power! Watt-hours (Wh) and kilowatt-hours (kWh) quantify energy consumed over a specific period of time. Power in watts or kilowatts quantifies the rate at which energy is expended or used at some moment in time. Each CFL saves you $60 - 12 = 48$ watts of power compared to the incandescent lamp at any given point in time. In 6 hours, that's 48×6, or 288 watt-hours (Wh) of consumed energy. If you have two dozen bulbs, you save $288 \times 24 = 6912$ Wh of consumed energy in the 6 hours per day that the lamps burn. A kilowatt-hour equals 1000 watt hours, so you save $6912/1000 = 6.912$ kWh per day. At 10 cents per kilowatt-hour, you save 69.12 cents ($0.6912) per day on the average. Over the course of a year, you'll save $0.6912 \times 365 = \$252.29$ (rounded off to the nearest penny). That sum will buy a pretty good supper for two in most American cities, or a great birthday party for your daughter or son!

Install Timers and Actuators

One of the oldest, and yet most effective, ways to conserve energy entails installing simple timers or motion-sensing actuators that can keep you from unnecessarily burning electric lights. These devices can, in some cases, also help you keep your home a bit more secure (when you're gone) than it would otherwise be.

Timers

Simple timers comprise small switch-equipped clocks that plug into an electrical wall outlet, and that have one or more outlets on them for use with electric lights and small appliances. You can set the timer to switch a lamp on at sundown, for example, and off at bedtime. Most of them contain simple mechanical clocks, although a few are entirely electronic and contain no moving parts.

If you have several of these devices connected to different lamps around the house, and if you set them to switch the lamps on and off at various times, you can go on a short vacation and uninformed neighbors will never suspect that you're away from home. You'll find these timers at nearly all major department stores and good hardware stores.

Here's a Tip!

You can use timers of the sort described above with small appliances, not only lamps. For example, you might have a timer device turn your TV set on at 7 p.m. and off at 11 p.m. every day when you're on vacation, and a second timer in another room switch a different TV set on at 8 p.m. and off at 10 p.m.

Motion-Sensing Light Switches

Motion sensors can detect the presence of moving objects and use the impulse to switch a light or other appliance on and off. The most common devices of this sort are housed in metal boxes with light-bulb sockets attached to them. Figure 8-1 is a photograph of a motion-sensing light switch intended for use with outdoor floodlights. It is housed in a cabinet roughly the size of an electrical switch box, and made from similar rugged metal. The lamp sockets contain rubber grommets to keep water from intruding and causing electrical problems.

When a large object comes into the range of the device and moves around, an electrical impulse goes to a delay switch, which actuates the lights for a certain period of time. If no further motion occurs, the lights stay on for just a minute or two; you can adjust the length of time before they automatically switch off. You can also adjust the sensitivity of the device, so it won't generate false positives when small insects approach it, yet will detect the presence of a moving person or large animal within a few feet. You can obtain switches of this type at most major department stores.

Did You Know?

Some motion-sensing light switches don't work well with CFLs or LEDs because, apparently, those lamps don't draw enough current to place a noticeable load on the semiconductor switches inside the sensors. The sensitivity and/or delay adjustments might act strangely or fail to work altogether. Outdoors, CFLs might not work during the winter in locations where the temperature reaches frigid extremes. You can experiment with your sensor and see how well CFLs and LEDs work with it. If your device doesn't "like" these lamp types, you should use incandescent lamps.

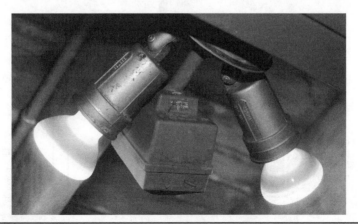

Figure 8-1 This photograph shows two 50-watt incandescent spot lights with a motion-sensing actuator between them. It's mounted under a deck, over the back entrance to a residential home.

Install a Heat Pump

A *heat pump* transfers thermal energy from one place to another. The term "pump" comes from the fact that the device uses a common external source of power, usually electricity, to move thermal energy rather than generating it directly.

Homogenize This!

All humanmade heating or cooling systems act against the natural process called *heat entropy* that takes place as temperatures gradually equalize throughout the universe. In a simplistic sense, all of our furnaces and air conditioners impose small-scale order in a gigantic thermodynamic system that relentlessly strives for chaos, a process whose ultimate endpoint some scientists have called the *heat death of the cosmos*.

Air-Exchange Heat Pump

Figure 8-2 shows the basic components of an *air-exchange heat pump*, also called an *air-source heat pump*, operating to transfer heat energy from the outdoor environment

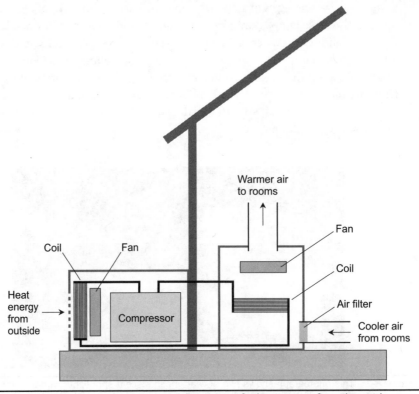

FIGURE 8-2 An air-source heat pump, operating to transfer heat energy from the outdoor environment into a building.

to the indoor environment (heating mode). The fan blows outdoor air through a coil that contains a refrigerant. As the refrigerant passes through the outdoor coil, depressurization and evaporation occur, causing the refrigerant to absorb heat energy. This process can occur even if the outdoor temperature is quite cool. The fluid then passes through pipes (shown as solid lines) to the indoor coil, where the refrigerant undergoes compression and condensation, causing it to release the heat energy it took in from the outside. The indoor coil becomes considerably warmer than ordinary room temperature; in fact, it can warm to as much as +35°C (+95°F) as the refrigerant passes through the indoor coil. The air that has been heated by the indoor coil then flows into the ductwork and circulates throughout the house.

Tip

Air-exchange heat pumps work well in the heating mode as long as the outdoor temperature remains above roughly +39°F (+4°C). If the outdoor air gets colder than that, it doesn't contain enough thermal energy to allow efficient operation.

Figure 8-3 portrays the same heat pump operating to transfer heat energy from the indoor environment to the outdoor environment (cooling mode). The fan blows

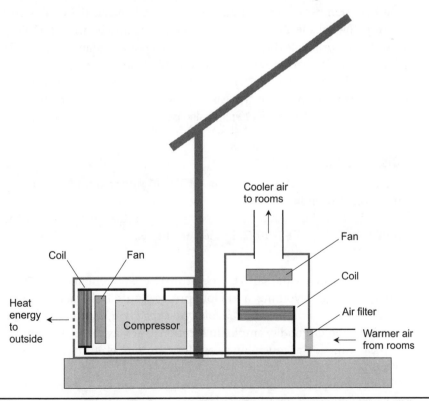

FIGURE 8-3   An air-source heat pump, operating to transfer heat energy from inside a building to the outdoor environment.

indoor air through a coil that contains a refrigerant. As the refrigerant circulates through the indoor coil, depressurization and evaporation occur, so the refrigerant absorbs heat energy, thereby chilling the air that moves past the coil. The cooled air then flows into the ductwork and circulates throughout the house. The chilling process can also remove some water vapor from the indoor air if the humidity is high, causing the indoor coil to "sweat." The heated fluid passes through pipes (shown as solid lines) to the outdoor coil. In the outdoor coil, the refrigerant undergoes compression and condensation, causing it to give up the heat energy that it acquired from indoors. The outdoor fan blows warm air into the external environment.

Did You Know?

Heat energy contained in the earth's atmosphere constitutes a renewable and practically unlimited resource.

Ground-Source Heat Pump

Some heat pumps extract thermal energy from beneath the earth's surface rather than from the outside air, and transfer this energy into a house or building. Figure 8-4 illustrates the principle. Basically, it's a modified air-exchange system. The outdoor coil resides underground, so the system requires no fan. In some arrangements, the outdoor coil can be placed near the bottom of a deep pond or lake. Heat transfer occurs by conduction from the earth to the coils. The system shown in Fig. 8-4 constitutes a *ground-source heat pump*, also called a *geothermal heat pump*.

Did You Know?

Ground-source heat pumps with sufficiently deep outdoor coil systems can function efficiently, even in places where winters grow severe. At a depth of several meters beneath the surface, the temperature remains constant all year round, at least +50°F (+10°C) in most locations.

In some locations, the earth temperature rises and remains high even at shallow depths. Saratoga, Wyoming, and Hot Springs, South Dakota, offer locations in the United States with plenty of available *geothermal heat* despite their severe winters. Iceland is another good example. In locations such as these, a ground-source heat pump can function at much lower outdoor air temperatures than an air-exchange heat pump can.

In locations where subsurface temperatures are high, even quite close to the surface, a network of pipes can replace the outdoor coil, buried just deep enough to

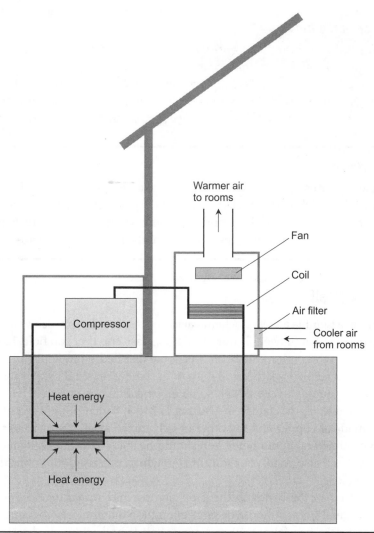

FIGURE 8-4   A ground-source (geothermal) heat pump, operating to transfer heat energy from the earth into a building.

allow heating of water that a mechanical pump (similar to the type of pump that brings water up from a well) can circulate through the house. That type of heating system is probably the most efficient design attainable in the real world! The only component that consumes any energy is the water pump, similar to the one in a well. Otherwise, you can literally get your heat for free (until your government finds a way to tax you for using it).

Ground-source heat pumps can cool the indoor environment during the summer in most locations. However, in some places, such as those mentioned above where the subsurface temperature is high, the cooling mode isn't practical.

For Nerds Only

A good heat pump can transfer more heat energy than it demands from the utility, so you'll sometimes hear a vendor claim that a heat pump's "efficiency" exceeds 100 percent. Actually, the vendor refers to the *coefficient of performance* (COP): the ratio of thermal energy transferred to the indoor environment to the input energy that the system needs to do the job.

Caveat!

Heat pumps, especially deep ground-source systems, cost a lot of money to install. You should expect several years to pass before you recover the up-front expense in terms of the month-to-month savings that you realize, compared with the cost of running the system that the heat pump replaced.

Build a Wet Cell

You can assemble an electrochemical *wet cell* and see how much voltage and current it produces. You'll need a short, fat, thick glass cup that can hold 12 ounces (about 0.36 liter) when full to the brim. Get some distilled white vinegar and table salt at your local supermarket. You'll need some bell wire (solid copper insulated wire of 18 to 22 gauge), a wire cutter, some electrical tape, some steel wool or an emery cloth, and two pipe clamps measuring 1/2 to 5/8 inch (1.3 to 1.6 centimeters) wide, one made of copper and the other of galvanized steel, designed to fit pipes 1 inch (2.5 centimeters) in diameter. You should be able to find these items at a hardware store. You'll also need your multimeter, which you used for applications earlier in this book.

Get rid of the bends in the pipe clamps, and straighten them out into strips. The original clamps should be large enough so that the flattened-out strips measure at least 4 inches (about 10 cm) long. Polish both sides of the strips with steel wool or a fine emery cloth to get rid of any *oxidation* that might have formed on the metal surfaces.

Cut two lengths of bell wire, each about 1.5 feet (50 centimeters) long. Strip 2 inches (5 centimeters) of insulation from each end of both lengths of wire. Pass one stripped wire end through one of the holes in the copper electrode, and wrap the wire around it two or three times, as shown in Fig. 8-5A. Wrap the other end of the stripped wire from the copper electrode several times around the positive (red) meter probe tip, as shown in Fig. 8-5B. Pass one stripped end of the other length of wire through one of the holes in the galvanized electrode, and wrap the other end of the wire around the negative (black) meter probe tip. Secure all four connections with electrical tape to insulate them and keep the wires from slipping off. Remove both of the meter probe leads from their receptacles on the meter.

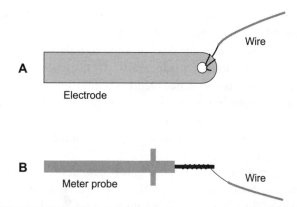

FIGURE 8-5 Attachment of wires to electrodes (at A) and meter probes (at B). Wrap the bare wire around the metal. Then secure the connections with electrical tape.

Lay the strips against the inner sides of the cup with their ends resting on the bottom. Be sure that the strips are on opposite sides of the cup so they're as far away from each other as possible. Bend the strips over the edges of the cup to hold them in place as shown in Fig. 8-6. Be careful not to break the glass! Fill the cup with vinegar until the liquid surface is slightly below the brim.

Once you've put the parts together, as shown in Figs. 8-5 and 8-6, add a rounded teaspoon of common table salt (sodium chloride). Stir the mixture until the salt completely dissolves in the vinegar. You'll know that all the salt has dissolved when you don't see any salt crystals on the bottom of the cup after you allow the liquid to stand still for a minute or two.

Set the meter to measure a low DC voltage. The best meter switch position is the one that indicates the smallest voltage that's greater than 1 volt. Insert the negative meter probe lead into its receptacle on the meter. Then insert the positive meter probe lead into its receptacle and note the voltage on the meter display. When I

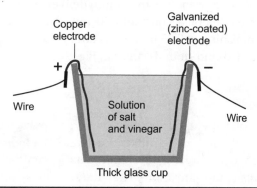

FIGURE 8-6 A wet cell made from a vinegar-and-salt solution. The glass cup has a brimful capacity of approximately 12 fluid ounces (0.36 liter).

conducted this experiment, I got a reading of 515 millivolts (or 0.515 volt). After 60 seconds, the voltage was still 515 millivolts.

Remove the positive meter lead from its receptacle on the meter. Set the meter for a low DC current range. The ideal setting is the lowest one showing a maximum current of 20 milliamps or more. Insert the disconnected meter lead back into its receptacle, and carefully note how the current varies with time. I got a reading of 8.30 milliamps to begin with. The current dropped rapidly at first, then more and more slowly. After 60 seconds, the current stabilized at 7.45 milliamps, as shown by the lowermost (solid) curve in Fig. 8-7.

When you conduct these tests, you'll probably get more or less voltage or current than I got, depending on how much vinegar you have poured into your cup, how strong the vinegar itself actually is, and how large your electrodes are. In any case, you should find that the open-circuit voltage remains constant as time passes, while the maximum deliverable current decreases.

Warning! If you get a notion to try any of these exercises with an automotive battery or other large commercial wet cell or battery, *forget about it!* The electrolyte (chemical solution that stores the energy) in that type of battery is a powerful acid that can boil out and burn you severely if you short-circuit the terminals. Some such batteries can even explode if you mistreat them, with obviously disastrous consequences.

Add another rounded teaspoon of salt to the vinegar. As before, stir the solution until the salt has completely dissolved. Repeat the voltage and current experiments. You should observe slightly higher voltages and currents. As before, the open-circuit voltage should remain constant over time, and the maximum deliverable current should fall. I measured a constant 528 millivolts. The current started out at 10.19 milliamps and declined to 8.76 milliamps after 60 seconds, as shown by the middle (dashed) curve in Fig. 8-7.

Add a third rounded teaspoon of salt and fully dissolve it. Once again, measure the open-circuit voltage and the maximum deliverable current. When I carried out this little exercise, I got a constant 540 millivolts. The current began at 11.13 milliamps, diminishing to 9.43 milliamps after 60 seconds passed, as shown by the uppermost (dashed-and-dotted) curve in Fig. 8-7.

Did You Know?

The higher voltage and current values with added salt result from increasing chemical activity of the electrolyte. If you add still more salt beyond the three rounded teaspoons already in solution, you'll eventually reach a point where the vinegar can't take any more. The solution will be *saturated,* having reached its greatest possible concentration.

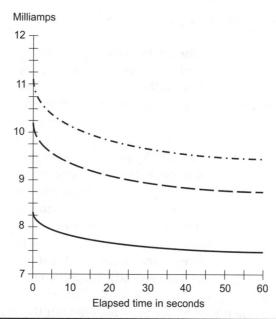

FIGURE 8-7   Graphs of maximum deliverable currents as functions of time for various amounts of salt dissolved in 12 fluid ounces of vinegar. Lower (solid) curve: 1 rounded teaspoon of salt. Middle (dashed) curve: 2 rounded teaspoons of salt. Upper (dashed-and-dotted) curve: 3 rounded teaspoons of salt.

When you leave the ammeter connected across the cell terminals for awhile, you'll notice that bubbles appear on the electrodes, especially with higher salt concentrations. The bubbles comprise gases (mainly hydrogen and oxygen, but also some chlorine) created as the electrolyte solution breaks down into its constituent elements. Although you won't see it in a short time, the electrodes will eventually become coated with solid material as well.

If you "short out" your wet cell and leave it alone for an extended period, all of the chemical energy in the electrolyte will ultimately get converted into heat. The maximum deliverable current will fall to zero, as will the open-circuit voltage. The cell will have met its demise, killed by its own juice!

Try This Experiment!

When you measure the voltage across the terminals of your wet cell without requiring that the cell deliver any current (other than the tiny amount required to activate the voltmeter), the cell doesn't have to do any work. You might expect that the voltage will remain constant for hours under those conditions. Let the cell sit idle overnight, with nothing connected to its terminals, and measure its voltage again tomorrow. What happens?

See How "Electric" You Are

In this experiment, you'll use a salt-and-vinegar solution to make contact between the electrodes and your hands, but most of the electrolyte will be inside your body! For this experiment, you'll need all the items left over from the previous experiment.

Remove the galvanized and copper electrodes from the vinegar-and-salt solution. Leave the solution in the cup. Leave the wires connected to the electrodes. Rinse the electrodes with water, dry them off, and get rid of the bends so they both form flat strips with holes in each end. Make sure that the probe leads are plugged into the meter. Then switch the meter to one of its most sensitive DC voltage ranges.

Wet your thumbs, index fingers, and middle fingers up to the first knuckles by sticking both hands into the vinegar-and-salt solution. (Don't be surprised if this solution stings your fingers a little bit. It's harmless!) Grasp the electrodes between your thumb and two fingers. Don't let your hands come into contact with the wires, but only with the metal faces of the electrodes. What does the meter say? When I conducted this experiment, I got a steady 515 millivolts.

Rinse your hands with water and dry them off. Switch the meter to its most sensitive DC current range. Wet your fingers with the vinegar-and-salt solution again, and grasp the electrodes in the same way as you did when you measured the voltage before. Watch the current level for 60 seconds, making sure that you don't change the way you hold onto the electrodes. The current reading should decline, rapidly at first, and then more slowly. When I did this experiment, the current started out at 122 microamps and declined to 95 microamps after 60 seconds had passed. Beyond 60 seconds, the current remained almost constant. Figure 8-8 illustrates the current graphed against time.

Set the meter to a different current range and repeat the above experiment. Don't expect to get the same readings as before. Of course, a small amount of variation is inevitable in any repeated experiment involving material objects. In this case, however, you should see a difference that's too great to be explained away by imperfections in the physical hardware.

An ideal ammeter would have no internal resistance, so it would have no effect on the behavior of a circuit when connected in series with that circuit. But in the real world, all ammeters have some internal resistance because the wire coils inside them don't conduct electricity perfectly. Unless it's specially engineered to exhibit a constant internal resistance, a meter set to measure small currents has a greater internal resistance than it does when set to measure larger currents. Most inexpensive test meters (such as mine) aren't engineered to get rid of these discrepancies.

When I changed the meter range while measuring my body current, my meter's internal resistance competed with my body's internal resistance. When I set the meter to a lower current range, I increased the total resistance in the circuit, reducing the actual flow of current. Conversely, as I set the meter to a higher current range, I decreased the total resistance in the circuit, increasing the actual current.

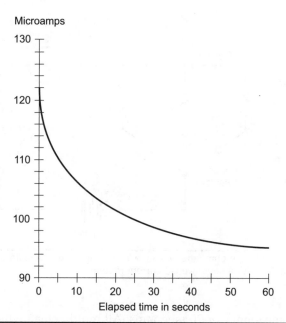

Microamps

FIGURE 8-8 Graph of maximum deliverable current as a function of time from my "body cell." I wetted my hands with a solution of 3 rounded teaspoons of salt dissolved in 12 fluid ounces of vinegar.

For Nerds Only

Does the foregoing phenomenon remind you of the *uncertainty principle* that physicists sometimes talk about? The actions of observers can change the behavior of the system that they look at! Sometimes, if not usually, this effect is too small to see. In this experiment, if you have the same or similar type of meter that I have, the uncertainty principle makes itself vividly apparent.

See How "Resistive" You Are

In the previous experiment, you discovered that your body resistance affects the amount of current that can flow in a circuit when your body forms part of that circuit. In this experiment, you'll measure your body resistance. You'll need everything you used in the previous experiment, along with a second copper electrode.

An *ohmmeter* (resistance-measuring meter) for DC actually comprises a milliammeter or microammeter in series with a set of fixed, switchable resistances and a battery that provides a known, constant voltage, as shown in Fig. 8-9. If the resistances are selected appropriately, the meter gives indications in ohms over any desired range. The device can be set to measure resistances from 0 ohms up to a certain maximum, such as 2 ohms, 20 ohms, 200 ohms, 2 kilohms, 20 kilohms, 200

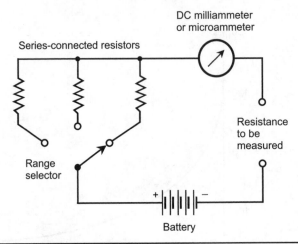

FIGURE 8-9   A multirange ohmmeter works by switching various resistors of known values in series with a sensitive DC current meter.

kilohms, 2 megohms, or 20 megohms. All multimeters, which you learned about earlier in this book, have an ohmmeter function.

Did You Know?

An ohmmeter must be calibrated at the factory, or in an electronics lab. A small error in the values of the series resistors can cause large errors in measured resistance. Therefore, these resistors must have precise *tolerances*. In other words, their values must actually be what the manufacturer claims, to within a fraction of 1 percent if possible. In addition, the battery must provide exactly the right voltage.

If you want to measure the resistance between two points with an ohmmeter, you must make sure that no voltage exists between the points where you intend to connect the meter. Such a preexisting voltage will add or subtract from the ohmmeter's internal battery voltage, producing a false reading. Sometimes, in this type of situation, an ohmmeter might say that a component's resistance is less than 0 ohms or more than "infinity" ohms!

Heads Up!

The measurement of internal body resistance is a tricky business. The results you get will depend on how well the electrodes are connected to your body, and also on where you connect them.

Get a second copper clamp from your local hardware store. Take the bends out of it to make it into a flat strip, and then polish it in the same way as you polished the other two electrodes. Connect one copper strip to each of the meter probe tips using bell wire. Switch the meter to measure a relatively high resistance range, say 0 to 20 kilohms. Dip your fingers into the electrolyte solution left over from the previous experiment. What does the meter say? Repeat the experiment using the next higher resistance range (in my meter, that would be 0 to 200 kilohms).

When I measured my body resistance using the above-described scheme, I got approximately 7.8 kilohms (that is, 7800 ohms) with the meter set for 0 to 20 kilohms, and 4.9 kilohms with the meter set for 0 to 200 kilohms. The difference resulted from internal meter resistance, just as in the previous experiment with the current-measuring apparatus. The higher resistance range required a different series of resistances than the lower range. These resistances appeared in series with the resistance of my body, so the total current flow (which is what the meter actually "sees") changed as the range switch position changed.

Fact or Myth?

One of my friends tried this experiment. He got 6.3 kilohms at the 0-to-20-kilohm meter range, and 4.5 kilohms at the 0-to-200-kilohm meter range. He wondered if the results of this experiment could serve as an indicator of a person's overall health. I said that I didn't think so, but that I really didn't know. I doubt that anyone can definitively answer that question one way or another!

Try this experiment with a copper electrode and a galvanized electrode, in the same arrangement as you used when you performed the previous experiment. Connect your body to the meter, as shown in Fig. 8-10A. Then reverse the polarity of your "body-electrode-meter" circuit by connecting the red wire to the black meter input, and connecting the black wire to the red meter input, so you get the configuration shown in Fig. 8-10B. You should observe different meter readings. You might even get a "negative" body resistance or a meter indication to the effect that the input is invalid.

For Nerds Only

Why do you think a discrepancy in the ohmmeter readings occurs when the electrode metals are dissimilar, but not when they're identical? Here's a hint: Imagine your body as a big bag of electrolyte solution!

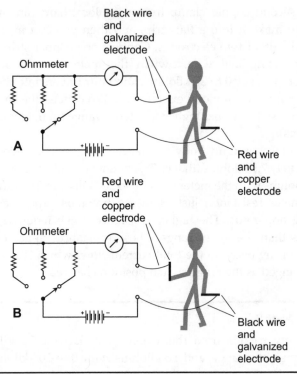

FIGURE 8-10   Try to measure your body resistance with the arrangement you used to measure current in the previous experiment, as shown at A. Then try the same test with the meter probe wires reversed, as shown at B.

Try Shortwave Radio

The range of radio frequencies from 3 to 30 MHz is sometimes called the *shortwave* band. Technically it's called the *high-frequency* (HF) range. The waves are actually long, as they travel through space, compared to the waves in most wireless communications these days. But they're short compared to the wavelengths that were most commonly used when the term was coined.

In the early 1900s, practically all communication and broadcasting was done at frequencies below 1.5 MHz (wavelengths of more than 200 meters). Engineers thought that the higher frequencies were useless. The vast region of the electromagnetic spectrum comprising wavelengths of "200 meters and down" (frequencies of 1.5 MHz and above) was given to amateur radio operators and experimenters, who became known as *hams*.

Within a few years, those hams discovered that the shortwave frequencies could support long-distance communications and broadcasting. In fact, the shortwave band worked better than the traditional *longwave band* did, allowing reliable contacts spanning thousands of miles using transmitters with only a few watts of output power. Soon, commercial entities and governments took keen interest in the shortwave band, and amateurs lost most of it. But the hams did manage to retain exclusive use of small slivers of the shortwave spectrum, and these so-called "HF ham bands" remain popular to this day.

Shortwave radio is still used for international broadcasting, particularly by developing countries. In the more technologically advanced nations, most government and business entities have moved their operations to the *very high frequencies* (VHF) from 30 to 300 MHz, the *ultra high frequencies* (UHF) from 300 MHz to 3 GHz, and the *microwave frequencies* above 3 GHz. This ongoing shift has given rise to talk of letting ham radio operators use more of the shortwave band that they originally discovered.

An HF radio communications receiver, especially one that offers continuous coverage of the HF spectrum, is sometimes called a *shortwave receiver*. Most general-coverage receivers function at all frequencies from 1.5 MHz through 30 MHz. Some also operate in the standard broadcast band at 535 kHz to 1.605 MHz. A few receivers, called *all-wave* receivers, can operate below 535 kHz, and into the so-called *longwave radio band*, as well as in bands at frequencies above 30 MHz.

Anyone can build or obtain a shortwave or general-coverage receiver, install a modest wire antenna in a backyard or even inside a frame house, and listen to signals from all around the world. This hobby is called *shortwave listening* (SWLing). Millions of people in the world enjoy it. In the United States, the proliferation of computers and online communications has largely overshadowed SWLing since the 1980s, and many young people grow up today ignorant of a realm of broadcasting and communications that still predominates in much of the world.

Various commercially manufactured shortwave receivers exist on the market, ranging in price from under $100 to thousands of dollars. A simple wire receiving antenna, which is all you need to receive the signals, costs practically nothing. Some of the better electronics or hobby stores carry these receivers, along with antenna equipment, for a complete installation. You can also shop around in consumer electronics and amateur radio magazines.

Did You Know?

A shortwave listener in the United States need not obtain a license to receive signals. In general, a license is required if shortwave transmission is contemplated. Shortwave listeners often get interested enough in communications to obtain amateur radio licenses, so they can engage in totally wireless conversations with other radio operators throughout the world, a mode that requires no supporting infrastructure other than the earth's upper atmosphere.

Try Amateur Radio

In most countries of the world, people can obtain government-issued licenses to send and receive messages via amateur radio. Hundreds of thousands of people have amateur radio licenses in the United States alone. Radio hams communicate by talking, sending Morse code, or typing on computer terminals. Typing the text on a computer resembles using the Internet. In fact, groups of amateur radio operators have set up their own radio networks, and some have "patched" into the Internet. Some hams, rather than talking or texting or using Morse code (also called "CW" for "continuous waves," even though technically they aren't continuous) on the radio, prefer to experiment with electronic circuits, and sometimes they come up with designs that later find their way into commercial equipment.

Some radio hams chat about anything they can think of except business matters, which are illegal to discuss using the ham radio frequencies in the United States. Others like to practice their emergency-communications skills, so they can be of public service during crises, such as hurricanes, earthquakes, or floods. Still others like to go out into the wilderness and talk to people thousands of miles away while sitting out under the stars. Radio hams communicate by radio from cars, trucks, trains, boats, aircraft, and even bicycles.

The simplest ham radio station has a transceiver (transmitter/receiver), a microphone, and an antenna. A small ham-radio station fits on a desk, and is about the size of a home computer or hi-fi stereo system. If you want, you can add accessories until your "rig" takes up an entire room in the house (or the better part of the basement, as in the author's installation shown in Fig. 8-11). You'll also need an antenna of some sort, preferably located outdoors. Figure 8-12 shows a simple vertical antenna designed to operate at 14 MHz, one of the most popular ham radio frequencies.

Amateur radio is an electronics-intensive and, increasingly, computer-intensive hobby. Radio hams are more likely to own two or more personal computers than are non-hams. Conversely, computer engineers and "power users" are more likely to be interested in ham radio than people who avoid computers. If you use a computer

FIGURE 8-11 The author's amateur radio station including a transceiver, Internet-ready computer, interface for digital radio communications, power meter, hi-fi audio system for big sound, and two displays, one of which is a flat-screen TV set.

FIGURE 8-12   The antenna for the ham radio station shown in Fig. 8-11. It's made of telescoping sections of aluminum tubing, mounted on a deck railing, and can be taken down or put up in a couple of minutes.

very much, and especially if you're interested in hardware design (as opposed to programming), you shouldn't have trouble obtaining an amateur radio license.

Figure 8-13 is a block diagram of a computerized ham radio station. The computer can be used to network with other hams who own computers, and it also serves as a terminal for the transceivers. The computer can also control the antennas

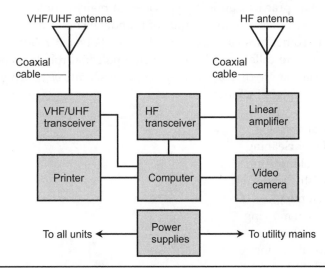

FIGURE 8-13   Block diagram of a sophisticated ham radio station. This design includes more peripherals than the station shown in Fig. 8-11, including a so-called "linear amplifier" for generating high-power signals at shortwave frequencies.

for the station, and can keep a log of all stations that have been contacted. Most modern transceivers can be remotely operated by computer over the radio or using the Internet.

For Aspiring Hams Only

You'll need to get a license to transmit on the amateur frequencies. The transmission of radio signals without a license is against the law, and can result in fines and/or imprisonment. There are several levels, or classes, of ham radio licenses available in the United States, all of which are issued by the Federal Communications Commission (FCC). For complete information, contact the headquarters of the *American Radio Relay League* (ARRL), 225 Main Street, Newington, CT 06111. This organization publishes books on all subjects relevant to ham radio, as well as license-exam study materials and a monthly magazine called *QST*, which means "Calling all radio amateurs." The people at ARRL headquarters can tell you the location of the nearest amateur radio club, where you can meet local hams and find out if this hobby is right for you. They maintain a website at www.arrl.org.

Get a Personal Robot

For centuries, people have imagined having *personal robots*. Until the explosion of electronic technology, however, people's attempts at robot-building resulted in clumsy masses of metal that did little or nothing of any real use. But now, personal robots can be practical, affordable devices for many people!

Personal robots can do all kinds of mundane chores around your house. Such robots are sometimes called *household robots*. Personal robots can be used in the office; these are called *service robots*. Some personal robots incorporate advanced features, such as *speech recognition*, *speech synthesis*, and *object recognition*. Household robot duties might include:

- Car washing
- Floor cleaning
- Cooking
- Dishwashing
- Laundry
- Lawn mowing
- Maintenance
- Meal serving
- Child's playmate
- Snow removal

- Toilet cleaning
- Window washing
- Companionship

Around the office, a service robot might do things such as:

- Floor cleaning
- Coffee preparation and serving
- Delivery
- Dictation
- Equipment maintenance
- Filing documents
- Greeting visitors
- Meal preparation
- Photocopying
- Toilet cleaning
- Window washing

Did You Know?

Some personal robots have been designed and sold. But until recently, they were not sophisticated enough to be of any practical benefit. Most such robots are more appropriately called *hobby robots*. Simpler machines make good toys for children. Interestingly, if a robot is designed and intended as a toy, it often sells better than if it is advertised as a practical machine.

Homogenize This!

You can enter "personal robot" (as a phrase) into your favorite Internet search engine, and you'll get some really cool hits. I found one place (as of this writing) called the Robot Shop (www.robotshop.com). Then I hit a link called "Domestic Robots" and got information about all sorts of machines, including vacuum cleaners, sweepers, companions, and even a robot golf caddy!

Table of Electricity at Utility Outlets in Various Countries

Voltages in the table are approximate (to within a few percent). Do not take these values as "the last word." Double-check with government authorities in the country of interest, and/or with a qualified travel agent, before purchasing your adapters. At the time of this writing, a comprehensive information guide for voltages and adapter plugs throughout the world was available at the website http://www.voltageconverters.com/voltageguide.htm.

Table of Electricity at Utility Outlets in Various Countries throughout the World as of Summer 2012 (Voltages are approximate—to within a few percent.)

Country	RMS Voltage	Line Frequency (in Hertz)
Australia	234	50
Bali	234	50
Barbados	117	50
Barbuda	234	60
Belarus	234	50
Belgium	234	50
Belize	117 or 234	60
Bermuda	117	60
Bhutan	234	50
Bolivia	117 or 234	50
Bosnia	234	50
Brazil	117 or 234	60
Bulgaria	234	50
Cambodia	117 or 234	50
Canada	117	60
Canary Islands	117	50
Cape Verde Islands	234	50
Cayman Islands	117	60
Chad	234	50
Channel Islands	234	50
Chile	234	50
China	234	50
Colombia	117	60
Congo	234	50
Costa Rica	117	60
Croatia	234	50
Cuba	117 or 234	60
Curacao	117	50
Cyprus	234	50
Czech Republic	234	50
Denmark	234	50
Dominica	234	50

Table of Electricity at Utility Outlets in Various Countries throughout the World as of Summer 2012 (Voltages are approximate—to within a few percent.) (*continued*)

Country	RMS Voltage	Line Frequency (in Hertz)
Dominican Republic	117	60
Ecuador	117	60
Egypt	234	50
El Salvador	117	60
England	234	50
Estonia	234	50
Ethiopia	234	50
Falkland Islands	234	50
Fiji	234	50
Finland	234	50
France	234	50
Gabon	234	50
Gambia	234	50
Georgia	234	50
Germany	234	50
Ghana	234	50
Greece	234	50
Greenland	234	50
Grenada	234	50
Grenadines	234	50
Guadeloupe	234	50
Guam	117	60
Guatemala	117	60
Guinea	234	50
Haiti	117	60
Honduras	117	60
Hong Kong	234	50
Hungary	234	50
Iceland	234	50
India	234	50
Indonesia	117 or 234	50
Iran	234	50

Table of Electricity at Utility Outlets in Various Countries throughout the World as of Summer 2012 (Voltages are approximate—to within a few percent.) (*continued*)

Country	RMS Voltage	Line Frequency (in Hertz)
Iraq	234	50
Ireland	234	50
Isle of Man	234	50
Israel	234	50
Italy	234	50
Jamaica	117	50
Japan	117	50 or 60
Jordan	234	50
Kazakhstan	234	50
Kenya	234	50
Kirghizstan	234	50
Korea	117 or 234	60
Malta	234	50
Mauritania	234	50
Mauritius	234	50
Mexico	117	60
Micronesia	117	60
Moldova	234	50
Monaco	117 or 234	50
Mongolia	234	50
Montenegro	234	50
Montserrat	234	60
Morocco	117 or 234	50
Mozambique	234	50
Myanmar	234	50
Namibia	234	50
Nepal	234	50
Netherlands	234	50
New Caledonia	234	50
New Zealand	234	50
Nicaragua	117	60
Niger	234	50

Table of Electricity at Utility Outlets in Various Countries throughout the World as of Summer 2012 (Voltages are approximate—to within a few percent.) (*continued*)

Country	RMS Voltage	Line Frequency (in Hertz)
Nigeria	234	50
Northern Ireland	234	50
Norway	234	50
Okinawa	117	60
Oman	234	50
Pakistan	234	50
Palau	117	60
Panama	117	60
Papua New Guinea	234	50
Paraguay	234	50
Peru	117 or 234	50
Philippines	117 or 234	60
Poland	234	50
Portugal	234	50
Puerto Rico	117	60
Qatar	234	50
Romania	234	50
Russia	234	50
Rwanda	234	50
Saudi Arabia	117 or 234	60
Scotland	234	50
Senegal	234	50
Serbia	234	50
Seychelles	234	50
Sierra Leone	234	50
Singapore	234	50
Slovak Republic	234	50
Slovenia	234	50
Solomon Islands	234	50
Somalia	117 or 234	50
South Africa	234	50
Spain	234	50

Table of Electricity at Utility Outlets in Various Countries throughout the World as of Summer 2012 (Voltages are approximate—to within a few percent.) (*continued*)

Country	RMS Voltage	Line Frequency (in Hertz)
St. Kitts / Nevis	234	60
St. Lucia	234	50
St. Martin	117 or 234	60 or 50
St. Vincent	234	50
Sri Lanka	234	50
Sudan	234	50
Suriname	117 or 234	60
Swaziland	234	50
Syria	234	50
Tazhikistan	234	50
Tahiti	117	60
Taiwan	117	60
Tanzania	234	50
Thailand	234	50
Tibet	234	50
Togo	117 or 234	50
Tonga	234	50
Trinidad and Tobago	117	60
Tunisia	117 or 234	50
Turkey	234	50
Turkmenistan	234	50
Tuvalu	234	50
Uganda	234	50
Ukraine	234	50
United Arab Emirates	234	50
United States	117	60
Uruguay	234	50
U.S. Virgin Islands	117	60
Uzbekistan	234	50
Vanuatu	234	50
Venezuela	117	50 or 60
Vietnam	117 or 234	50

Table Electricity at Utility Outlets in Various Countries throughout the World as of Summer 2012 (Voltages are approximate—to within a few percent.) (*continued*)

Country	RMS Voltage	Line Frequency (in Hertz)
Wales	234	50
Western Samoa	234	50
Yemen	234	50
Yugoslavia	234	50
Zambia	234	50
Zimbabwe	234	50

Glossary

Alternating current—Abbreviation, AC. Electrical current that reverses direction at regular intervals. In the United States and some other countries, standard utility AC reverses direction every 1/120 of a second, so it goes through 60 complete cycles per second. In many countries, standard utility AC reverses direction every 1/100 of a second, so it goes through 50 complete cycles per second.

Alternator—A small electric generator used in motor vehicles to provide power to electrical and electronic devices, and to keep the battery charged up. The alternator in a car or light truck can generate several amperes of current when the engine runs at normal driving speed.

Alternator whine—Interference to mobile two-way radio reception caused by the electrical activity in the vehicle's alternator. Usually sounds like an audio "tone" whose pitch varies depending on the speed of the engine. Sometimes this whine can appear on a transmitted signal, too.

Amateur radio—Also called "ham radio." A popular worldwide electronics hobby. In most countries of the world, people need government-issued licenses to operate ham radio transmitters. Some operators communicate over wireless media at various frequencies by talking, sending Morse code, or typing on computer terminals. Others prefer to experiment with electronic circuits. Of course, any individual "ham operator" can do both of these things, as well as contribute to emergency communications preparedness.

Ampere—The standard unit of electrical *current*. Mathematically it can be considered the equivalent of a *volt* per *ohm* in a simple DC circuit.

Analog television—Also known as *fast-scan TV* (FSTV) or *National Television System Committee* (NTSC) TV. In most of the world, broadcasters no longer use this mode; it was pretty much done away with worldwide by 2011. Nevertheless, if you have an old TV set, it was probably designed for analog TV and won't work nowadays unless you get a digital-to-analog converter box.

Audio mixer—A hi-fi sound system component that allows you to connect the outputs of multiple audio devices to a single channel input for an amplifier. The mixer isolates the amplifier's inputs from each other, so you don't have to worry about any possible mismatch or "competition" among the source devices. In addition, you can adjust the signal level (gain) for each device without affecting the behavior of any other device.

Automotive battery—A large, heavy battery about half the size of a concrete cinder block. In the United States, a standard automotive battery produces 12.6 V at full charge with no load. It contains six 2.1-V *lead-acid cells* connected in series.

Baffle—An object, usually a flat piece of wood or other sound-reflective material, used to modify the way that sound waves propagate in a speaker cabinet, room, or auditorium. Properly used, baffles can enhance the quality of sound from a high-end audio system.

Balance control—A single control, or a pair of controls, intended for adjusting the relative volume levels of the sounds coming from the left and right channels in a stereo hi-fi system. This control can compensate for variations in speaker placement, relative loudness in the channels, and the acoustical characteristics of the room in which the equipment is installed.

Banana connector—A convenient single-pin plug-and-receptacle combination. Used with single-wire conductors to make temporary connections.

Bar code tag—A simple *possession-based access control* technology. Also used for item labeling. The tag has parallel bands of various widths. More sophisticated tags have complicated patterns of black shapes on a white background. A laser rapidly scans the pattern. The dark regions absorb the laser light, while the white regions reflect the light back to a sensor. The sensor, thereby, receives a binary data signal unique to the pattern on the tag.

Battery—A combination of electricity-producing *cells* that produces a higher *voltage*, or that can deliver a higher *current*, than an individual cell can do all by itself.

Binary digital signal—Signals that occur in either of two states called *bits* (a contraction of "binary digits") that represent the number 1 and the number 0, but nothing else. These signals are, in fact, rapidly fluctuating direct currents. Commonly used in computer communications.

Biometric access control—A property protection technology that acts on certain biological characteristics of people authorized to enter. For example, it might employ a camera along with a pattern recognition computer program to check a person's facial contours against information in a gigantic database. The machine might use speech recognition to identify people by breaking down the waveforms of their voices. It might record a hand print, fingerprint, or iris print. It might even employ a combination of all these things. A computer analyzes the data obtained by the sensors, and determines whether or not the person has authorization to enter the premises.

Bit—A single element in a *binary digital signal*. It can exist in either of two states, but only those two states, called 1 (high) and 0 (low).

Breaker—See *Circuit breaker*.

Cassegrain dish—A satellite TV or satellite Internet dish antenna whose geometry resembles that of a Schmidt-Cassegrain reflecting telescope. Sometimes found in more remote areas where a larger antenna is necessary. This type of dish can sometimes measure more than two meters in diameter. The signal arrives along the dish axis, reflects from the spherical or paraboloidal surface, and comes to a focus at a second, smaller reflector. The second reflector causes the incoming *microwaves* to travel straight back to the center of the dish, where the energy enters the feed horn and frequency converter assembly through a small hole.

Catastrophic failure—An equipment malfunction in which the entire device or system becomes useless. Often takes place suddenly, and can happen without warning.

Cell—A device that produces DC electricity from chemical reactions, visible light radiation, or other processes. A cell can't be broken down into anything more elementary; it's the simplest possible arrangement for producing DC electricity. Compare *Battery*. Sometimes a cell is imprecisely called a "battery."

CFL—See *Compact fluorescent lamp*.

Charge controller—In an alternative electric power system with batteries, an essential component that makes sure the battery bank receives the optimum amount of charging current at all times from the solar panel, wind turbine, water turbine, or other energy-producing device. Also prevents overcharging, which can shorten the life of a *deep-cycle battery*.

Charge polarity—The "sense" of electricity, either negative (an excess of electrons) or positive (a shortage of electrons). Sometimes negative is called "minus" and positive is called "plus." In electric circuits, electrons flow from regions having relatively more negative polarity to regions having relatively more positive polarity. Physicists consider the current to flow from positive to negative in theory, but that's an old convention that scientists invented before they knew that electrons, which carry negative charge, existed.

Chip—See *Integrated circuit*.

Circuit breaker—A current-actuated switch that performs the same basic functions as a *fuse*, but that can be reset, and does not need to be replaced every time it "trips." Most homes in the United States use circuit breakers in their electrical systems. Some individual appliances, such as electric space heaters and *transient* *suppressors*, also contain circuit breakers, typically rated at 15 A in the United States (for 117-V circuits).

Clip lead—A short length of flexible wire, equipped at one or both ends with a simple, temporary connector.

Coaxial cable—A single-conductor *shielded cable* in which a length of wire, called the center conductor, is surrounded by a cylinder of solid metal, a cylindrical wire braid, or a layer of metal foil called the outer conductor or shield. The center conductor is separated from the shield by a layer of dielectric (insulating) material such as polyethylene. Commonly used in radio and television communications systems, as well as for high-speed Internet connections.

Coefficient of performance—An expression of how well a *heat pump* works. Sometimes mistakenly called "heat-pump efficiency." Mathematically, the COP equals the ratio of thermal energy that the system transfers to the amount of energy that the system consumes in order to do its job.

Combustion generator—An independent source of electricity that uses an internal combustion engine. Most small generator engines burn gasoline. Larger ones burn diesel fuel, propane, or methane.

Compact fluorescent lamp—Abbreviation, CFL. A lamp or bulb comprising a coiled-up, miniaturized fluorescent tube with electronic circuits that allow it to function much more efficiently than traditional *incandescent lamps*. You can expect a CFL to last far longer than an incandescent lamp in the same application. Some concerns have arisen about CFLs because they contain mercury, a known environmental toxin. Nevertheless, CFLs have attained popularity as direct replacements for incandescent lamps.

Compact hi-fi system—The simplest type of home audio arrangement. It resides in a single cabinet. The speakers can be either internal or external; if they're external, the connecting cables are short. The assets of a compact system include small size, simplicity, and low cost.

Component hi-fi system—A home audio arrangement with dedicated equipment cabinets containing components that perform specific functions. The individual units are interconnected with cables, ideally of the shielded coaxial type. A component system costs more than a compact system does, but you get better sound fidelity, more audio power, the ability to do more tasks, and the opportunity to tailor the system to your preferences.

Conductance—The ease with which an electrical *current* can flow through a substance.

Conductor—A substance in which electrical *current* can flow easily. Examples include most metals, particularly copper and aluminum, which are used for making wire and in the construction of wireless antennas.

Continuity test—In an electrical circuit or system, a test that you can conduct with a *multimeter* to find out whether two points are directly connected or not. If they are, then you will observe no DC or AC voltage between them, and the resistance between them will equal zero.

Conventional dish—A satellite TV or satellite Internet dish in which the signal arrives at a slight angle with respect to the dish axis, reflects from the spherical or paraboloidal metal surface of the dish, and then enters a device called a feed horn, which acts like an "ear for *microwaves*." The feed horn is connected to a converter that changes the frequency of the signal so that it can travel along a coaxial cable to the TV equipment inside your house. The whole assembly measures less than 1 meter wide, 1 meter long, and 1 meter deep.

Conversion efficiency—For a *photovoltaic cell*, the ratio of the available electrical output power to the total radiant power striking the cell (with both quantities expressed in the same units, such as watts). Can be multiplied by 100 to obtain the figure as a percentage.

Corona—A visible glow that commonly occurs when extreme voltages on electrical conductors ionize the surrounding air. A corona can appear at the ends of a radio-transmitting antenna element. It is also seen in high-tension power lines. A corona occurs increasingly often as the relative humidity rises because it takes less voltage to ionize moist air than it takes to ionize dry air. A corona is normally not visible in daylight, but it shows up clearly at night, when it is sometimes called "Saint Elmo's fire."

Current—The movement of particles that carry an electrostatic charge from one point to another. Usually these particles are electrons.

Decibel—Abbreviation, dB. A real-world unit of relative sound loudness, based on the way people perceive it. If you change the volume control on a hi-fi set so that you can just barely tell the difference in the loudness when you anticipate the change, then that change equals approximately 1 dB. Decibels are worked out by engineers with respect to the threshold of hearing, which represents the faintest sound that you can detect in an otherwise silent place, assuming that you have good hearing.

Degree of phase—See *Phase*.

Deep-cycle battery—A rechargeable battery that can store a lot of energy, and that you can repeatedly charge and discharge to run electrical devices. Often used with alternative electric power systems so that they get their original energy from renewable, natural sources, such as the sun, the wind, or moving water.

Digital logic—A form of "reasoning" used by electronic machines, particularly devices and systems controlled by computer chips. Works on the basis of two conditions, called high (usually logic 1) and low (usually logic 0). These conditions are like "truth" and "falsity" in mathematical logic.

Digital spread spectrum—Abbreviation, DSS. A communications mode in which the signal frequency hops or sweeps over a defined range. The transmitter frequency varies according to a specific, encoded, repeating pattern. No receiver can hear the signal unless it "knows the code" and acts on it correctly.

Dip—A momentary decrease in the power-line voltage that can occur when a large appliance first comes on. You've probably noticed these fluctuations as momentary "blinks" in old-fashioned incandescent bulbs when some heavy appliance, such as a washing machine or refrigerator, starts up.

Direct current—Abbreviation, DC. Electrical *current* that always flows in the same direction. The intensity (or strength) of the current can vary, but in true DC, the polarity (direction) never changes. Nearly all electronic equipment needs DC in order to function properly. That's why, unless they operate from batteries, your gadgets and systems usually have specialized power supplies.

Dirty electricity—A term that describes the fact that ordinary household utility electricity contains energy at many frequencies other than the 60 Hz "main" AC frequency (or 50 Hz in much of the world outside the United States). All of the frequency components produce *electromagnetic fields* that can sometimes interfere with wireless electronic devices.

D-shell connector—A specialized, multiconductor plug and jack combination with trapezoidal (D-shaped) shells that force you to insert the plug correctly into the jack.

Dynamic range—The difference in *decibels* between the strongest and the weakest output audio signals that the system can produce without objectionable distortion. It's a prime consideration in hi-fi recording and reproduction. As the dynamic range specification of an amplifier increases, the sound quality improves for music or programming having a wide range of volume levels.

Dynamic speaker—A coil-and-magnet combination that translates alternating electrical *current* into mechanical vibration, thereby producing sound waves in the air. Used in most hi-fi sound systems.

Electric eye—An intrusion-detection system with a visible-light or infrared (IR) source, usually a laser diode, and a sensor, such as a photoelectric or photovoltaic cell. These devices are connected into an actuating circuit. When something interrupts the light or IR beam, the *current* or *voltage* passing through, or generated by, the sensor changes. An electronic circuit detects this change and sends a signal to an alarm or other alerting device.

Electric field—A "region of influence" that occurs in space around all electrically charged particles and objects. Electric fields are responsible for the attraction and repulsion that we observe between charged objects.

Electric flux lines—Imaginary contours in space that define the orientation, direction, and intensity of an *electric field*.

Electrical ground—A connection in the wiring of a home or business that remains at zero *voltage* with respect to the earth. In any good home or business electrical system, the electrical ground connection is established at the point where the utility lines enter the building, usually in the form of a ground rod at or near the distribution panel (fuse or breaker box). In a well-engineered electrical system, the "third prong" in every wall outlet should be connected directly to the building's electrical ground.

Electrolyte—In an electrochemical cell or battery (such as the zinc-carbon, alkaline, lead-acid, or lithium type), the solution or paste inside the component. This solution or paste contains energy in chemical form, which converts to electrical energy when you connect a lamp, radio, tablet computer, or other electronic device to the cell or battery.

Electromagnet—A magnet constructed with a coil of wire around a "magnetizable" rod of metal, such as iron or steel. When current flows through the coil, the rod becomes a magnet. When current stops flowing through the coil, the rod loses its magnetism.

Electromagnetic (EM) field—A complex energy field comprising an alternating or fluctuating *electric field* and an alternating or fluctuating *magnetic field* working together, arising from *alternating current* in a wire or from certain physical phenomena. Radio waves, infrared, visible light, ultraviolet, x rays, and gamma rays all manifest themselves as EM fields.

Electromagnetic (EM) spectrum—The range of all *electromagnetic field* (EM field) frequencies or wavelengths commonly encountered in the real world. Frequencies range from a few hertz to millions of terahertz; wavelengths range from many kilometers down to a tiny fraction of a millimeter.

Electrostatic charge—An excess or shortage of electrons that tends to persist, so that an object maintains a positive or negative *charge polarity* with respect to its surroundings. Sometimes lay people call this condition "static electricity."

Electrostatic speaker—A pair of large, flat, closely spaced metal plates that translates electrical voltage into mechanical vibration, thereby producing sound waves in the air. Used in some high-end hi-fi audio systems.

Ethernet cable—A cable that connects a *modem* or a hard-wired *router* to your computer for Internet use. It looks like a telephone landline cord with similar plugs on the ends, but the Ethernet cable has more wires than an old-fashioned phone cord does.

Female connector—A receptacle, such as a wall outlet, into which a *male connector* fits.

Firewall—A computer program or hardware device that helps to keep rogue computer wizards (known as *hackers* in technical jargon) from taking control of your computer when it's connected to the Internet. No firewall is perfect, however; the best hackers can get through them all.

Frequency—The number of complete cycles that occur every second in *alternating current* (AC) or pulsating *direct current* (DC).

Fuel cell—A device that converts combustible gaseous or liquid fuel into electricity at a lower temperature than normal combustion does. Hydrogen is a common energy source for these devices. The hydrogen oxidizes to form energy and water (along with a small amount of nitrous oxide if air serves as the oxidizer). The proton exchange membrane (PEM) fuel cell represents one of the most widely used technologies. A single PEM hydrogen fuel cell generates approximately 0.7 V DC. When two or more fuel cells are connected in series to get higher voltage, the combination is called a stack.

Fuse—An electrical component that protects circuits and devices from overload. Found in older houses, and in most motor vehicles. Also found in some individual electrical appliances. In a motor vehicle, the fuses protect the battery and the alternator in case of a short circuit. Fuses also minimize the risk of electrical overheating that can cause fires. Vehicle headlights, brake lights, turn signals, backup lights, interior lights, climate-control fans, radios, and other electrical devices all have fuses in their lines. If a fuse blows, the affected device or circuit won't get the electricity it needs, so it can't work.

Generator efficiency—In an electric generator, the ratio of the electrical power output to the mechanical driving power, both measured in the same units, such as watts (W) or kilowatts (kW), and multiplied by 100 to get a percentage.

Global Positioning System—Abbreviation, GPS. A network of radiolocation and radionavigation devices that operates on a worldwide basis. The GPS employs numerous satellites, and allows you to determine your location on the earth's surface, and in some cases, your altitude above the surface as well.

Graphic equalizer—A device that lets you adjust the relative loudness of audio signals at various frequencies. It allows for meticulous tailoring of the sound quality in hi-fi equipment. The circuit contains several independent gain controls, each one affecting a different part of the audible spectrum. Acts as a sophisticated *tone control*.

Grid-intertie system—See *Interactive system*.

Hacker—A smart computer user who likes to break into other people's systems, usually over the Internet. Some hackers simply enjoy the challenge and the intrigue of snooping around in other people's business, and the technical process can be fascinating. More nefarious hackers break into systems for the purpose of stealing people's money, identity, or other information. Still others do it to make a political statement.

Ham radio—Common slang for *amateur radio*.

Harmonic—A signal or wave that exists at a whole-number multiple of the main or fundamental frequency. For example, if 60 Hz is the fundamental frequency, then harmonics can exist at 120 Hz, 180 Hz, 240 Hz, and so on. The 120 Hz wave forms the second harmonic (twice the fundamental frequency), the 180 Hz wave forms the third harmonic (three times the fundamental frequency), the 240 Hz wave forms the fourth harmonic (four times the fundamental frequency), and so on.

Head unit—In a mobile sound system, the component that gathers signals from the radio, and/or converts data from media, such as compact-disc (CD) players, MP3 players, or tape cassettes, into audio signals. When you buy a vehicle, it will have a factory-installed head unit in the dashboard, where both the driver and the front-seat passenger can reach it.

Headset—Also known as a pair of headphones. A hi-fi component that offers listening privacy, keeps your "big sound" experience from disturbing people around you, and gets rid of sound-wave reflection problems inherent in all systems that use speakers. In effect, a headset comprises two small dynamic speakers, one placed directly against (or very close to) each ear.

Heat pump—An appliance that transfers thermal energy from one place to another to warm up or cool down the indoor environment. The term "pump" comes from the fact that the system uses a common external source of power, usually electricity, to move thermal energy rather than generating it directly.

Hermaphroditic connector—An electrical plug/jack pair with two or more contacts, some of them male and some of them female. In some cases, hermaphroditic connectors at opposite ends of a single length of cable look identical when viewed "face-on." However, the pins and holes have a special geometry, so you can join the two connectors in the correct way only.

High-Definition Multimedia Interface—Abbreviation, HDMI. On a big-screen display or computer monitor, a port and cable that will let you connect your set to an up-to-date computer so that you can view programs on the Internet, and also look at homemade videos that you can create using popular devices, such as webcams and camcorders.

High-definition television—Abbreviation, HDTV. Any of several methods for getting more detail into a TV picture than could ever be done with *analog television*. The HDTV mode also offers superior sound quality, making for a more satisfying home TV and home theater experience. High-definition TV is transmitted in a digital mode. Digital signals propagate better than analog signals do, they're easier to deal with when they are weak, and they can be processed in ways that analog signals would not allow.

High frequencies—Abbreviation, HF. The range of radio frequencies from 3 MHz up to 30 MHz, corresponding to wavelengths from 100 meters down to 10 meters. Also called the *shortwave band*.

Hydrogen fuel cell—See *Fuel cell*.

Ignition noise—Interference to radio reception caused by internal combustion engines. Typically sounds like a rapid sequence of "pops" or a steady "buzz." A special form of *impulse noise*.

Impulse noise—Interference to radio reception caused by a sudden, high-amplitude voltage pulse or regular sequence of pulses. This noise can come from all kinds of household appliances, such as vacuum cleaners, hair dryers, electric blankets, thermostats, and fluorescent-light starters. Impulse noise tends to get worse as the frequency goes down, and can plague AM broadcast receivers to the consternation of their users.

Incandescent lamp—A lamp (or bulb) that works by allowing an electric current to flow through a piece of wire that has a precisely tailored resistance and current-carrying capacity. As a result, the wire, called the filament, glows white hot.

Infrared—The portion of the *electromagnetic spectrum* with wavelengths longer than those of visible light but shorter than those of radio *microwaves*.

Infrared motion detector—An intrusion alarm system that uses *infrared* (IR) devices. Two or three wide-angle IR pulses are transmitted at regular intervals; these pulses cover most of the room in which the device is installed. A sensor picks up the returned IR energy, normally reflected from the walls, the floor, the ceiling, and the furniture. The intensity of the received pulses is measured and recorded by a microprocessor. If anything in the room changes position, the intensity of the received energy will vary, and the resulting signal will set off an alarm.

Infrared presence detector—An intrusion alarm system that employs a simple IR sensor, in conjunction with a microprocessor, to detect sudden increases in the amount of IR (or "radiant heat") present in a particular place. The time threshold can be set so that gradual changes, such as might be caused by the sun warming a room, do not trigger the alarm, while rapid changes, such as a person entering the room, will set it off.

Insulator—A substance in which electric *current* cannot flow under normal circumstances. Examples include most plastics, glass, dry wood, paper, and dry air.

Integrated circuit—Abbreviation, IC. A "wafer" of semiconductor material with many components etched onto it to create a device that performs a specific function in an electronic system. Computers are built largely with these devices. Also called a chip.

Interactive system—In solar, wind, and hydroelectric alternative energy use, a system that connects with your commercial electric utility provider's circuits to reduce or offset your electric bill. Compare *Stand-alone system*.

Inverter—See *Power inverter*.

Kilowatt hour—Abbreviation, kWh. The amount of electrical energy that a 1000-watt (1-kilowatt) appliance consumes in an hour, or the equivalent of it.

Knowledge-based access control—A property-protection technology in which people are issued numerical codes. The entrances to your property have locks that disengage when the proper sequence of numbers is punched into a keypad. This keypad can be hard-wired into the system, or it can be housed in a box about the size of a cell phone. It works like a bank automatic-teller machine (ATM) personal identification code.

Lamp cord—Also known as "zip cord" and consisting of two stranded wires embedded in rubber or plastic insulation. Commonly used with small appliances, such as table lamps, portable battery chargers, and clock radios.

Lead-acid cell—A common type of rechargeable electrochemical cell, used in automotive and power-backup applications. When the cell has a full charge, the negative electrode consists of pure lead, and the positive electrode consists of lead dioxide. The electrolyte, which contains all the battery's energy in chemical form (and which converts to electrical energy when something demands current), is sulfuric acid diluted with water.

LED—See *Light-emitting diode*.

Light-emitting diode—Abbreviation, LED. A *semiconductor* device in the form of a lamp (or bulb) that is specially designed to emit visible light when current passes through it. These lamps produce a reasonable amount of light but consume only a small amount of electrical power. As a result, LEDs cost less to operate, once you buy them, than any other type of lamp known as of this writing.

Logic—See *Digital logic*.

Logic gate—A device, usually etched onto an *integrated circuit*, that performs a specific function in *digital logic*.

Log-periodic antenna—Also called a log-periodic dipole array (LPDA). For TV and FM broadcast reception in remote areas where cable service is not available and satellite service is not desired, an antenna that consists of several straight, parallel metal elements called dipoles, all of which are connected together with a pair of wires. The shortest dipole is closest to the feed point that connects to the ribbon or cable going down to your receiver. The longest element resides at the back of the antenna, farthest from the feed point.

Magnetic field—A "region of influence" that occurs in space whenever charged particles, such as electrons, move, or when the atoms in certain substances align with each other. All magnets are surrounded by these "regions of influence." Magnetic fields are responsible for the attraction and repulsion that we observe between magnets in close proximity.

Magnetic flux lines—Imaginary contours or "threads" in space that define the orientation, direction, and intensity of a *magnetic field*.

Male connector—A connector with exposed prongs or pins, such as the plug on the end of a lamp cord, that fits into a receptacle called a *female connector*.

Memory drain—A phenomenon that can occur with nickel-based rechargeable cells and batteries, especially the older nickel-cadmium (NICAD) types. The device loses its ability to deliver current after only a partial discharge, so that the useful life between recharging sessions is greatly reduced. In some cases, this problem can be overcome by discharging and recharging the device several times. Memory drain does not occur with lithium-based rechargeable cells and batteries, which have largely replaced the nickel-based ones in recent years.

Microwave frequencies—Also called simply microwaves. In its broadest context, the range of radio frequencies above 300 MHz, corresponding to wavelengths shorter than 1 meter. The literature disagrees about the exact upper and lower limits of the microwave band: some texts put the lower limit at 1 GHz (a wavelength of 30 centimeters) and others put it at 3 GHz (a wavelength of 10 centimeters). In no case, however, does this range extend into the infrared (IR) portion of the spectrum, which begins at frequencies of around 1 THz (a wavelength of 0.3 millimeter or 300 micrometers) and ends at the lower limit of the visible-light range.

Midrange speaker—A speaker that handles sound at frequencies near the middle of the audio range. Commonly combined with a *tweeter* and a *woofer* in high-end hi-fi sound systems.

Modem—A device that encodes signals going out from your computer into the Internet, and decodes signals coming in from the Internet to your computer. A modem can link a computer to the same cable system that provides your TV service. Some modems are designed to connect directly to a network of optical fibers. Still others contain a small radio transceiver for wireless or satellite access. The most primitive work with a telephone landline to get you a dial-up connection.

Multimeter—A simple laboratory instrument designed to measure current, voltage, and resistance in electrical circuits. Also called a *volt-ohm-milliammeter* (VOM).

Negative-ground vehicle—A motor vehicle in which the negative battery terminal goes directly to the metal frame or chassis, which represents the reference voltage level, called "common" or "ground." This arrangement is found in the vast majority of cars and trucks in the United States.

Ohm—The standard unit of electrical *resistance*. Mathematically it can be considered the equivalent of a *volt* per *ampere* in a simple DC circuit.

Ohmic loss—The loss of power or energy that occurs in a long span of electrical transmission line, as a result of *resistance* in the wires.

Ohmmeter—A meter designed to measure electrical *resistance*.

Parallel connection—A method of combining two or more electrical components by joining all their left-hand ends together and all their right-hand ends together, getting an arrangement in which the components resemble the rungs in a ladder, and the interconnecting wires resemble the two vertical supports in the ladder. In a parallel circuit, all the components receive the same voltage, but the currents through them might differ. Compare *Series connection*.

Parametric equalizer—A hi-fi audio *graphic equalizer* in which the gain (volume), center frequency, bandwidth, and skirt slopes are independently adjustable. In addition to several bandpass filters for each channel, a parametric equalizer can incorporate a low-frequency *shelf filter* and a high-frequency shelf filter, both of which have adjustable gain, shelf frequency, and skirt slope. It's the most sophisticated form of *tone control* commonly available.

Passive transponder—A wireless *possession-based access control* system. It's a magnetic tag that authorized people can wear or carry. They're the same little things that department stores employ to deter petty thieves. The transponder can be read from several feet away.

Phase—A technical term that refers to points along an AC wave cycle. Phase can also express the extent of the timing difference between two AC waves that have the same *frequency*. Often expressed in degrees, where one degree of phase equals 1/360 of a cycle.

Phone jack—A *female connector* with recessed contacts that mate with *phone plug* contacts. The contacts have built-in spring action that holds the phone plug in place after insertion. The most common diameters are 1/4 inch (6.35 millimeters) and 1/8 inch (3.175 millimeters).

Phone plug—A *male connector* with a rod-shaped metal sleeve and one or two other contacts. A ring of hard-plastic insulation separates the contacts. The most common diameters are 1/4 inch (6.35 millimeters) and 1/8 inch (3.175 millimeters).

Phono connector—A simple plug or jack designed for coaxial cable at low voltages and low current levels. You simply push the plug onto the jack, or pull it off. Also known as an *RCA connector*. Commonly used with audio equipment.

Photovoltaic cell—Abbreviation, PV cell. An electronic component that converts visible light, infrared, or ultraviolet rays directly into DC electricity. When used to obtain electricity from sunlight, this type of device is known as a solar cell. A silicon PV cell produces approximately 0.5 V DC in sunlight. Multiple PV cells can be connected together to get higher voltages in so-called PV panels, PV batteries, PV modules, and PV arrays.

PIN lock—Acronym for "personal identification number lock." A setting that makes it practically impossible for anyone to use your cell-phone set if you lose it. You should activate the PIN lock if you want to keep your account, and any personal information that you might have stored on the phone, secure. Experts recommend that everybody who has a cell phone "lock it down" with a PIN code right away after buying it.

Pixel—A contraction of the words "picture element." It's the smallest unit of visible information in a video image or display. Each pixel can have any of numerous hues (color tints), saturation (color richness) levels, and brightness (actual brilliance) levels, independently of all the other pixels. A video display will carry a specification that tells you the number of pixels going horizontally. Some will tell you both the horizontal and vertical values.

Possession-based access control—A property-protection method that requires authorized people to possess a physical object that unlocks the entry to your property. Magnetic cards are a popular form of possession-based security devices. You insert the card into a slot, and a microcomputer reads data encoded on a magnetic strip. This data can be as simple as an access code, of the sort you punch on a keypad.

Power inverter—An electronic device that produces utility AC, usually 117 V at 60 Hz (in the United States), from low-voltage DC, usually 12 to 24 V. The best power inverters synthesize a *sine wave* resembling the output at a typical household utility outlet.

Power supply—A circuit that converts utility AC into pure DC suitable for use with electronic devices. Many devices have their power supplies built in. Each particular electronic device requires DC at a specific voltage, and also needs a certain amount of current.

Precipitation noise—Also called precipitation static. Radio interference caused by electrically charged water droplets or ice crystals as they strike metallic objects, especially antennas. The resulting discharge produces wideband noise that sounds similar to the noise generated by electric motors, fluorescent lights, or other appliances.

Proton exchange membrane—Abbreviation, PEM. See *Fuel cell*.

Rack-mounted hi-fi system—A home audio ensemble with multiple hardware units, similar to a *component hi-fi system*, but with all the units built to a single, standardized width for installation in a vertical rack. The rack can be mounted on wheels so that you can easily move the whole system, except for external speakers, from place to place.

RCA connector—See *Phono connector*.

Rectifier—A device or circuit that converts AC to pulsating DC. Usually comprises one, two, or four semiconductor diodes.

Relay—An electromechanical device that allows for remote switching of large appliances. A small electromagnet causes a lever to open and close the circuit.

Resistance—The extent to which a substance opposes the flow of *current*.

RMS voltage—An expression of the effective or DC-equivalent voltage in an AC circuit. The abbreviation RMS stands for "root mean square," a mathematical process that engineers use to define the effective or DC-equivalent voltage. In most countries, the RMS voltage of standard utility AC electricity is either 117 V or 234 V, give or take a few percent.

Quick charger—A device that replenishes the charge in an electrochemical cell or battery by driving a relatively large current through it for a relatively short period of time. Not recommended. Compare *Trickle charger*.

Router—A device that allows you to access a single Internet connection with more than one computer (although you can use a router even if you have only one computer). Routers come in two types: hard-wired and wireless. To use a router, you plug it into your modem in place of a computer, activate the router according to the instruction manual, and then access the Internet from your computer(s) through the router and the modem combined.

Semiconductor—A substance in which electric current can flow easily under certain conditions, and with difficulty (or not at all) under other conditions. Examples include certain solid elements and compounds including silicon, selenium, germanium, gallium arsenide, and the oxides of some metals. The conductance can be controlled to generate, amplify, modify, mix, rectify, and switch electrical currents or electronic signals.

Series connection—A method of combining two or more electrical components by joining them end-to-end, getting an arrangement that geometrically resembles the links in a chain. In a series circuit, all the components carry the same current, but the voltages across them might differ. Compare *Parallel connection*.

Sferics—Radio-frequency noise generated in the atmosphere of our planet, mostly by lightning discharges in thundershowers. In a radio receiver, sferics cause a faint background hiss or roar, punctuated by bursts of sound called "static." You can hear sferics on an AM broadcast receiver during the summer in regions where thundershowers commonly occur.

Shelf filter—A device that allows you to adjust the volume of an audio amplifier circuit as a function of frequency. A low-frequency shelf filter keeps the volume constant above a certain critical frequency; below that frequency, the volume is adjustable. A high-frequency shelf filter keeps the volume constant below a certain critical frequency; above that frequency, the volume is adjustable. Shelf filters act like sophisticated *tone controls*. However, they're not as sophisticated as *graphic equalizers*.

Shielded cable—An electrical or electronic cable surrounded by a cylindrical wire braid, solid metal cylinder, or layer of metal foil that's connected to ground. The shield keeps *electromagnetic* (EM) *fields* from getting into or out of the cable.

Shortwave band—The range of radio frequencies from 3 to 30 MHz. Technically known as *high frequencies*. The waves are actually long, as they travel through space, compared to the waves in most wireless communications. But they're short compared to the wavelengths that were most commonly used when the term was coined in the early 1900s.

Shortwave listening—Abbreviation, SWLing. A popular electronics hobby in which people listen to signals from all around the world on the *shortwave band*. In the United States, the proliferation of computers and online communications has largely overshadowed SWLing since the 1980s, and many young people grow up today ignorant of a realm of broadcasting and communications that still predominates in much of the world.

Sine wave—An electrical waveform in *alternating current* (AC), in which all of the energy exists at a single *frequency*. This type of wave gets its name because, if viewed on a laboratory oscilloscope, it looks like a graph of the mathematical sine function. Also called a sinusoid.

Single-phase AC—Utility electricity that consists of a single, pure AC sine wave. You'll find it at standard wall outlets intended for small appliances, such as lamps, TV sets, and computers.

Smart electric meter—A sophisticated electric utility meter that has no moving parts. This type of meter can record various details about your electric energy consumption, such as peak power demand and changes in usage patterns. It can also interconnect with supplemental systems, such as solar panels and wind turbines, allowing you to reduce your monthly electric bill and, in some cases, actually profit by selling surplus power to the electric company.

Solar cell—See *Photovoltaic cell*.

Split-phase AC—Utility electricity that consists of two AC sine waves that travel along their own dedicated wires, with a third wire connected to electrical ground. The two waves directly oppose each other in *phase*.

Stack—See *Fuel cell*.

Stand-alone system—In solar, wind, and hydroelectric alternative energy use, a system that works all by itself, independent from commercial electric utility providers. Compare *Interactive system*.

Surge—The initial high current drawn by a poorly designed power supply when it's first switched on with a load connected, or a momentary increase in power-line voltage that lasts longer than a transient but is less intense. Don't confuse this term with *transient*.

Surge protector—See *Transient suppressor*.

Three-phase AC—Utility electricity in the form of three sine waves, each having the same *voltage*, but differing in relative *phase* by 120° (1/3 of a cycle). Each wave travels along its own wire, so the transmission line has three wires.

Three-wire electrical system—In most homes and buildings, an electrical wiring scheme that includes three wires: "live," "neutral," and "ground." This type of system requires the installation of three-slot wall outlets. In order to work effectively, the "third slot" (usually D-shaped and situated below the two vertical slots) in each outlet must go directly to the building's *electrical ground*.

Tone control—A control found in most audio amplifiers and also in many radios, CD players, and MP3 players. Allows the listener to adjust the relative bass (low-frequency) and treble (high-frequency) sound to get the best audio quality. In its simplest form, a tone control consists of a single rotatable knob or linear-motion sliding control. The counterclockwise, lower, or left-hand settings of this control result in strong bass and weak treble audio output. The clockwise, upper, or right-hand settings result in weak bass and strong treble.

Transceiver—A wireless transmitter and receiver contained in a single package. Examples include all cell phone sets, most ham radios, and all CB radios.

Transformer—A device that increases or decreases the AC voltage in an electrical system. A step-up transformer increases the voltage; a step-down transformer decreases the voltage.

Transient—A voltage "spike" on the AC utility line that can greatly exceed the positive or negative peak AC voltage. It usually lasts only a few millionths of a second, but if it's extreme, it can cause damage to sensitive electronic equipment. Don't confuse this term with *surge*.

Transient suppressor—A device that eliminates *transients* on an AC power line. Most such devices can be plugged into a standard wall outlet, and the protected equipment plugged into the suppressor box, which might contain up to half a dozen individual outlets. Sometimes inaccurately called a "surge protector."

Trickle charger—A device that replenishes the charge in an electrochemical cell or battery by driving a small current through it for a long time. This type of charger works better than a so-called *quick charger*.

Tuner—A home hi-fi stereo system component that contains a radio receiver. A typical tuner can receive signals in the standard AM broadcast band and/or the standard FM broadcast band. Some tuners can also receive satellite radio signals if you have a subscription to a service of that sort. Tuners don't have built-in amplifiers. A tuner can provide enough power to drive a headset, but you'll probably want to add an "outboard" amplifier to provide sufficient power for a pair of speakers.

Tweeter—A speaker designed especially for enhanced treble reproduction, and found in most high-end audio systems. Commonly used in conjunction with a *midrange speaker* and a *woofer*.

Twist splice—A simple, temporary wire splice, made by twisting the ends of two wires together and then putting electrical tape over the connection.

Two-wire electrical system—In older homes and buildings, an electrical wiring scheme that includes only two wires, one "neutral" and the other "live." You can recognize this type of system by the presence of two-slot outlets in the walls.

Ultra high frequencies—Abbreviation, UHF. The range of radio frequencies from 300 MHz up to 3 GHz, corresponding to wavelengths from 1 meter down to 10 centimeters.

Ultrasonic motion detector—An intrusion alarm system that detects motion by sensing changes in ultrasonic waves, which are acoustic waves at frequencies above the range of human hearing. A set of transducers sends out ultrasonic waves. Another set of transducers picks up the waves reflected from objects in the room. If anything shifts position, the relative *phase* of the waves will change. This data goes to a microprocessor, which can trigger an alarm.

Uninterruptible power supply—Abbreviation, UPS. A temporary backup power source with a battery that charges from the AC utility under normal conditions, but provides a few minutes of emergency AC if the utility power fails. That time allows you to deploy a backup generator without having to shut any of the connected devices down, or else shut down your computers and other devices properly before removing power altogether.

Vacuum tube—An amplifying device found in older electronic systems (mainly before 1970). Largely obsolete, these components needed high voltages in order to function, and they consumed power vastly out of proportion to the actual work that they did. Vacuum tubes are still used today in some high-power radio transmitters and high-fidelity audio amplifiers.

Valve—British expression for *vacuum tube*.

Very high frequencies—Abbreviation, VHF. The range of radio frequencies from 30 MHz up to 300 MHz, corresponding to wavelengths from 10 meters down to 1 meter.

Video Graphics Array—Abbreviation, VGA. On a big-screen display or computer monitor, a port and cable that will let you connect your set to a computer (even an old one) for viewing images.

Volt—The standard unit of electrical *voltage*. Mathematically, volts equal *amperes* times *ohms* in a simple DC circuit.

Voltage—Also called potential difference or electromotive force. An expression of the "pressure" caused by a charge difference between two points in a circuit, causing *current* to flow.

Volt-ohm-milliammeter—See *Multimeter*.

Volume-unit meter—Abbreviation, VU meter. Also called a distortion meter; used in high-end audio amplifier systems. Excessive input causes the meter needles or bars to "kick up" into the red range of the scales during audio peaks. You should operate your amplifiers so that the VU meter readings stay below the red range.

Waveform—The shape of a sound wave, radio wave, AC wave, or other signal wave as it would look on the screen of a laboratory oscilloscope. With this instrument, you get a graph of voltage versus time, with time on a horizontal axis going from left to right, and voltage on a vertical axis where downward represents negative and upward represents positive.

Western Union splice—A method of wire splicing that involves twisting the ends of two wires together from opposite directions, and then covering the entire splice with electrical tape. Solder can be applied before covering the splice to get the best possible electrical and mechanical bond.

Wi-Fi hotspot—The zone near a wireless *router* connected to an Internet *modem*, such as the one you get as part of a cable or satellite installation. You'll find Wi-Fi hotspots in public libraries, hotels, motels, restaurants, bars, and airports. Even a few fast-food places and department stores have them. You can bring your notebook computer or Wi-Fi-equipped tablet device into such a place, obtain the password from one of the employees, and get on the Internet.

Wind turbine—An electromechanical system that converts wind energy to electric energy. The most common type has a large propeller, resembling the ones in old-fashioned aircraft, attached to an electric generator with a rotatable shaft. However, other geometries exist, and new ones are constantly being developed.

Woofer—A speaker designed especially to reproduce low-frequency sound. Commonly found along with a *midrange speaker* and a *tweeter* in a single cabinet.

Zip cord—See *Lamp cord*.

Suggested Additional Reading

Chiras, Daniel D., *The Homeowner's Guide to Renewable Energy: Achieving Energy Independence through Solar, Wind, Biomass, and Hydropower*, 2nd ed. Gabriola Island, BC, Canada: New Society Publishers, 2011.

Chiras, Daniel D., *The Solar House: Passive Heating and Cooling*. White River Jct., VT: Chelsea Green, 2002.

Davidson, Homer, *Troubleshooting & Repairing Consumer Electronics without a Schematic*, 3rd ed. New York: McGraw-Hill, 2004.

DeGunther, Rik, *Alternative Energy for Dummies*. Hoboken, NJ: Wiley Publishing, Inc., 2009.

Ewing, Rex A., and Pratt, Doug, *Got Sun? Go Solar*, 2nd ed. Masonville, CO: PixyJack Press, 2009.

Frenzel, Louis E., Jr., *Electronics Explained*. Burlington, MA: Newnes/Elsevier, 2010.

Geier, Michael, *How to Diagnose and Fix Everything Electronic*. New York: McGraw-Hill, 2011.

Gerrish, Howard, *Electricity and Electronics*. Tinley Park, IL: Goodheart-Wilcox Co., 2008.

Gibilisco, Stan, *Electricity Demystified*, 2nd ed. New York: McGraw-Hill, 2012.

Gibilisco, Stan, *Electronics Demystified*, 2nd ed. New York: McGraw-Hill, 2011.

Gibilisco, Stan, *Teach Yourself Electricity and Electronics*, 5th ed. New York: McGraw-Hill, 2011.

Gipe, Paul, *Wind Energy Basics*, 2nd ed. White River Jct., VT: Chelsea Green, 2009.

Gussow, Milton, *Schaum's Outline of Basic Electricity*, 2nd ed. New York: McGraw-Hill, 2009.

Hodge, B. K., *Alternative Energy Systems*. Hoboken, NJ: Wiley Publishing, Inc., 2009.

Horn, Delton, *Basic Electronics Theory with Experiments and Projects*, 4th ed. New York: McGraw-Hill, 1994.

Horn, Delton, *How to Test Almost Everything Electronic*, 3rd ed. New York: McGraw-Hill, 1993.

Kybett, Harry, *All New Electronics Self-Teaching Guide*, 3rd ed. Hoboken, NJ: Wiley Publishing, Inc., 2008.

MacKay, David, *Sustainable Energy without the Hot Air*. Cambridge, UK: UIT Cambridge Ltd., 2009.

Miller, Rex, and Miller, Mark, *Electronics the Easy Way*, 4th ed. Hauppauge, NY: Barron's Educational Series, 2002.

Mims, Forrest M., *Getting Started in Electronics*. Niles, IL: Master Publishing, 2003.

Morrison, Ralph, *Electricity: A Self-Teaching Guide*, 3rd ed. Hoboken, NJ: Wiley Publishing, Inc., 2003.

Pittman, Aspen, *The Tube Amp Book*. San Francisco, CA: Backbeat Books, 2003.

Shamieh, Cathleen, and McComb, Gordon, *Electronics for Dummies*, 2nd ed. Hoboken, NJ: Wiley Publishing, Inc., 2009.

Slone, G. Randy, *TAB Electronics Guide to Understanding Electricity and Electronics*, 2nd ed. New York: McGraw-Hill, 2000.

Yoder, Andrew, *Auto Audio*, 2nd ed. New York: McGraw-Hill, 2000.

Index